ADAM
=ALIEN

VOL. 1.2

⋆⫯ LEON BIBI ⫯⋆

"Men would rather believe than know."

--E.O. Wilson

DEDICATION

I dedicate this book to the true, bold masters of our origins, through their dedication to tell the real story, despite the cynics. Their research and evidence is overwhelming, grip my imagination and convinced me that my gut was right—Zecharia Sitchin, Freidman, Pye, Hatcher Childress, Kasten, Thomas, Komarek, Roberts, Conway, Bauval, McDowell, Tellinger, Farrell, La Violette, and of course, Erich Von Daniken.

I also dedicate this to my family—my wife, Stacey; sons, Matthew and Zachary; and my daughter, Rebecca—for their love, their inspiration, and their support.

May you open your mind to a new belief system based on credible data and physical evidence. My vision of the world changed drastically after reading these brilliant authors' works. They absorb me in their thorough investigation of history, evolution, science, and religion. My hope is that my work inspires you, the reader, to question the origin of us. To read my work, believe it is true, and speak of it with conviction!

This book contains a secret, the topics of which have been forbidden from human history for 4000 years.

TIMELINE

DATE	EVENT
443,000 BC -	Arrival of the Anunnaki
442,000 BC -	360,000 BC - Anu comes to Earth
415,000 BC -	Ninhursag established her Medical center in Shurupak
335,000 BC -	Ice Age
226,983 BC -	Enki moves to Africa to supervise mining
220,000 BC -	Ice Age
183,783 BC -	Rebellion of the Anunnaki mine workers
180,000 BC -	Creation of Homo Sapiens
176,583 -	Garden of "E.DIN" - Adam and Eve procreate
127,000 BC -	Ice Age
20,983 BC -	Noah is born with a "genetically engineered skin color"
20,880 BC -	Noah has 3 children- Shem, Ham and Japheth
12,364 BC -	Nibiru enters Earth's orbit and forces Earth to tilt on axis
11,600 BC -	Ice Age
10,983 BC -	The Flood - Antarctic Ice Sheet slips into Indian Ocean. Massive tsunami overwhelms the Arabian Peninsula and floods the Persian Gulf
10,450 BC -	The Pyramids of Giza are built
8,764 BC -	Nibiru returns to Earth's orbit
8,700 BC -	Jerusalem built as a space facility. The Sphinx is carved
8,670 BC -	Second Pyramid War ends
5,164 BC -	Nibiru returns to Earth's orbit
4000 BC -	Uruk becomes the largest city in Mesopotamia. Anu visits Earth again.
3800 BC -	Sumerians write down first recorded history on clay tablets, in Uruk
3760 BC -	The Jewish calendar begins
3450 BC -	Nimrod builds the Tower of Babel for Marduk, and it is destroyed by Enlil

TIMELINE

DATE	EVENT
3100 BC -	Pharaonic dynasties begin
3000 BC -	Stonehenge is built by Thoth for Marduk as a star-clock
2700 BC -	Enlil resides in Nippur
2500 BC -	Uruk has a population of 40,000 people
2123 BC -	Birth of Abraham
2024 BC -	Anunnaki set off a nuclear weapon at Sodom and Gomorrah The Dead Sea bears its name. Many Anunnaki leave the earth
2024 BC -	Famine and hardship from the blast ends Sumerian civilization
2001 BC -	Jacob is born
1992 BC -	Abraham dies
1513 BC -	Moses discovered in the river
1450 BC -	Mohenjo-Daro and Harappa in current day Afghanistan destroyed by a nuclear bomb
1393 BC -	Yahweh created by Israelites as the one true God is documented in the Old Testament
1391 BC-	Teotihuacan pyramids are built
1308 BC -	Israelites exodus out of Egypt
968 BC -	King Solomon born
946 BC -	King Solomon builds Temple of Jerusalem for Yahweh
925 BC -	Temple of Jerusalem is destroyed by Ramses the Great
610 BC -	The balance of the Anunnaki left Earth
586 BC -	Nebuchadnezzar burns King Solomon Temple to Enlil
500 BC -	Old Testament was written
200 BC -	Anunnaki leave Earth for good

TABLE OF CONTENTS

CHAPTER 1
UFOS

I can assure you the flying saucers, given that they exist, are not constructed by any power on earth.

--President Harry S . Truman

I occasionally think how quickly our {US and Russia} differences worldwide would vanish, if we were facing an alien threat from outside this world.

--President Ronald Reagan

The phenomenon of UFO's does exist, and it must be treated seriously.

--Mikhail Gorbachev, former Soviet Premier

A few insiders know the truth... and are studying the bodies that have been discovered.

--Astronaut Edgar Mitchell

Some people asked, you know, were you alone out there? We never gave the real answer, and yet we see things out there, strange things, but we know what saw out there. And we couldn't really say anything. The bosses were really afraid of this, they were afraid of the War of the Worlds type stuff, and about panic in the streets. So we had to keep quiet.

--Astronaut John Glenn

I've never seen one, but I've convinced their real . I look up in the sky constantly, dying to see one. I may have seen one driving on Highway 1 in North Miami Florida, about a year ago, but I'm still not sure. Both my brothers have seen one. My youngest saw five UFO's over the Atlantic Ocean in 1989 during the summer in East Hampton. He claims they were with "like burning, burning deep red balls of fire, hovering over the water." My oldest saw the Hudson County UFO in 1981 off highway 684 in Westchester, New York. He swears he saw a huge ship in the shape of a triangle, possibly three football fields long, moving silently at around 6:00 p.m. He pulled over in his car on the shoulder of the highway, and another man was there pointing and looking at the object. My brother said to the man, "We're really not alone, are we?" The other man nodded. There were seven thousand witnesses of the Hudson Valley UFOs. I repeat, seven thousand! This is not a random incident, not an environmental mirage. This is real.

Here are some facts that lend support to the credibility of life outside of earth

- There are about one billion stars in our galaxy.
- There are about ten billion planets.
- If we exclude frozen planets and planets sterilized by heat, we estimate that there are about a minimum of ten thousand planets in our galaxy alone that can sustain life.
- Earth is one of those ten thousand.
- The nearest star is one light year away--fairly close in the scheme of planetary travel.

- Scientists have discovered how to bend the space-time continuum.

Here are some points that lend support to the existence of UFOs:

- If UFOs didn't exist, were a figment of the human imagination, pose no threat to national security, and existed only in the imagination, why then did the Air Force spend hundreds of

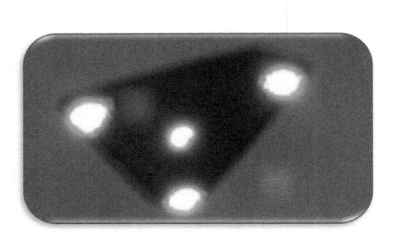

Triangular UFO over Belgium—visible Magnetic fields
seen, and confirmed by experts.

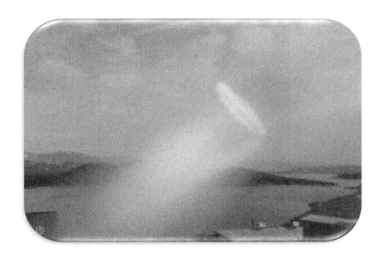

Norway 1957. This is not a light mirage. Photo confirmed and supported by Dr. J. Allen Hynek.

millions of dollars to create Project Blue Book (formally Project Sign and Project Grudge), the Condon Committee, APRO (Aerial Phenomena Research Organization), and NICAP (National Investigative Committee for Aerial Phenomenon)?

- Commander Sergeant Major Robert O. Dean, who served in the Korean War, said that the United States had observed on radar in 1960 approximately fifty UFOs, flying in formation, emerge from Russia, fly over Europe, and then turn toward the North Pole. An assessment was written between 1961 and 1964 that covered a history of UFO phenomena over Europe. The conclusions of the study were startling! The authors determined, **with certainty**, that we have been actively visited by at least four extraterrestrial civilizations for thousands of years! The conclusion of the report stated, "Evidence collected and studied in this report indicates that there is some kind of process or plan unfolding…and this survey [of earth and humanity] has been going on for a very long time; possibly thousands of years. There [did] not appear to be a major military threat involved…but if they were either hostile or malevolent there was absolutely nothing that we could do" (Kasten). According to Dean, Great Britian, the United States, and Germany did most of the research, but Air Marshall Sir Thomas Pike, former chief of the Royal Air Force, wrote it. Only 15 copies of this report were produced.

- According to the book *The Day After Roswell*, Colonel Phillip J. Corso, stated the following technological innovations were reversed engineered *directly* from remnants of the Roswell crash: Night vision technology, fiber optics, lasers, and transistors.
- A secret memo dated September 23, 1947, written by General Nathan Twinning, commander of the Army Air Force's Air Material Command at Wright Field, states the following:
  As requested by AC/AS-2 there is presented below the considered option of this Command concerning the so-called "Flying Discs"...

a. This phenomenon reported is something real and not visionary or fictitious.

b. There are objects probably the shape of a disk, of such appreciable size as to appear to be as large as man-made aircraft.

c. The reported operating characteristics such as extreme rates of climb, maneuverability (particularly in role), and action which must be considered evasive when sighted or contacted by friendly aircraft and radar, lend belief to the possibility that some of the objects are controlled either manually, automatically or remotely.

- Admiral Roscoe Hillenkoetter, the first director of the CIA, said, "behind the scenes, high ranking Air Force officers are soberly concerned about UFOs. But through official secrecy and ridicule, many citizens are led to believe the unknown flying objects are nonsense." (Komarek)

- Wilbert B. Smith, with the Department of Transportation in Canada, had a conversation with Dr. Robert Sarbacher (MJ-12) in September 1950 that went as follows (according to Smiths handwritten notes):

Smith: I have read Scully's book on the saucers and would like to know how much of it is true.

Sarbacher: the facts reported in the book are substantially correct.

Smith: Then the saucers do exist?

Sarbacher: Yes, they exist.

Smith: Do they operate as Scully suggests, on magnetic principles?

Sarbacher: We have not been able to duplicate their performance.

Smith: Do they come from another planet?

Sarbacher: All we know is that we did not make them, and it's pretty certain they did not originate from earth. Smith: I understand the whole subject is classified.

Sarbacher: Yes, it is classified two points higher even than the H-bomb. In fact, it is the most highly classified subject in the US government at the present time.

Smith: May I ask the reason for the classification?

Sarbacher: You may ask but I can't tell you.

(La Violette)

The following countries have disclosed their classified UFO Data: France (2007), New Zealand (2008) and the UK (2011). The following countries have not: United States, China, and Russia.

APPENDIX F

SECRET GOVERNMENT MEMOS CONCERNING
OPERATION MAJESTIC TWELVE

Below is a reproduction of Dr. Robert Sarbacher's Canadian Department of Transportation memo describing the U.S. effort to reverse-engineer UFO technology.

DEPARTMENT OF TRANSPORT
INTRA-DEPARTMENTAL CORRESPONDENCE

"Smith" memo.

Geo-Magnetics

CONFIDENTIAL

OTTAWA, Ontario, November 21 1950.

(R.ST.)

MEMORANDUM TO THE CONTROLLER OF TELECOMMUNICATIONS:

For the past several years we have been engaged in the study of various aspects of radio wave propagation. The vagaries of this phenomenon have led us into the fields of auroral, cosmic radiation, atmospheric radio-activity and geo-magnetics. In the case of geo-magnetics our investigations have contributed little to our knowledge of radio wave propagation as yet, but nevertheless have indicated several avenues of investigation which may well be explored with profit. For example, we are on the track of a means whereby the potential energy of the earth's magnetic field may be abstracted and used.

On the basis of theoretical considerations a small and very crude experimental unit was constructed approximately a year ago and tested in our Standards Laboratory. The tests were essentially successful in that sufficient energy was abstracted from the earth's field to operate a voltmeter, approximately 50 millivolts, although this unit was far from optimum design, it demonstrated the correctness of the basic principles in a qualitative manner and provided useful data for the design of a better unit.

The design has now been completed for a unit which should be self-sustaining and in addition provide a small surplus of power. Such a unit, in addition to functioning as a 'pilot power plant' should be large enough to permit the study of the various reaction forces which are expected to develop.

We believe that we are on the track of something which may well prove to be the introduction to a new technology. The existence of a different technology is borne out by the investigations which are being carried on at the present time in relation to flying saucers.

While in Washington attending the MARS Conference, two books were released, one titled "Behind the Flying Saucer" by Frank Scully, and the other "The Flying Saucers are Real" by Donald Keyhoe. Both books dealt mostly with the sightings of unidentified objects and both books claim that flying objects were of extra-terrestrial origin and might well be space ships

25

from another planet. Scully claimed that the preliminary studies of
one saucer which fell into the hands of the United States Government
indicated that they operated on some hitherto unknown magnetic
principles. It appeared to me that our own work in geo-magnetics
might well be the linkage between our technology and the technology
by which the saucers are designed and operated. If it is assumed that
our geo-magnetic investigations are in the right direction, the theory
of operation of the saucers becomes quite straightforward, with all
observed features explained qualitatively and quantitatively.

I made discreet enquiries through the Canadian Embassy
staff in Washington who were able to obtain for me the following
information:

a. The matter is the most highly classified subject in the United
   States Government, rating higher even than the H-bomb.

b. Flying saucers exist.

c. Their modus operandi is unknown but concentrated effort is being
   made by a small group headed by Doctor Vannevar Bush.

d. The entire matter is considered by the United States authorities
   to be of tremendous significance.

I was further informed that the United States authorities are investigating
along quite a number of lines which might possibly be related to the saucers
such as mental phenomena and I gather that they are not doing too well since
they indicated that if Canada is doing anything at all in geo-magnetics they
would welcome a discussion with suitably accredited Canadians.

While I am not yet in a position to say that we have solved
even the first problems in geo-magnetic energy release, I feel what the
correlation between our basic theory and the available information on
saucers checks too closely to be mere coincidence. It is my honest opinion
that we are on the right track and are fairly close to at least some of the
answers.

Mr. Wright, Defence Research Board liaison officer at the
Canadian Embassy in Washington, was extremely anxious for me to get in touch
with Doctor Solandt, Chairman of the Defence Research Board, to discuss with
him future investigations along the line of geo-magnetic energy release.

•••••••• 3

THE FONTES BRIEFING

Dr. Olavo Fontes, a prominent Brazilian UFO researcher in the 1950's, was given a briefing by two American intelligence agents whom he met on February 27th,1958, that stated the following:

1. All major world governments know about the existence of flying saucers.
2. Six flying discs have already crashed on earth and have been dismantled.
   a. Three crashed in the United States, two in Europe, and one in the Sahara desert.
   b. All six craft were smaller than 100 feet in diameter.
   c. There were dead bodies found in all of them.
   d. All the bodies were under 46 inches in height.
   e. All the bodies were "humanoid" in appearance.

3. All discs were in the shape of a saucer with a cabin on top.

4. The disks had the following characteristics:
   a. They were all made of a light metal.
   b. They had portholes made of an unknown type of glass.

5. All discs were propelled by a powerful electromagnetic field.
   a. The fields were a rotating, oscillating a high voltage electromagnetic type.

b. The larger ships had some form of atomic engines.

He goes on to say that the whole matter has been concealed from the public, that a carefully planned censorship has been in place for several years, and that the policy is to debunk the whole subject of flying saucers and ridicule anybody that claims to see one. He also says that force, sometimes deadly, is used to silence anyone that does not comply.

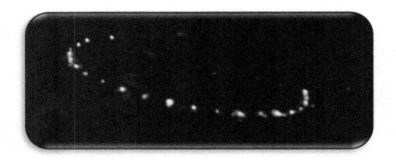

Hudson Valley UFO

One of the first ever UFO photos that has never been
discredited. Oregon 1950's.

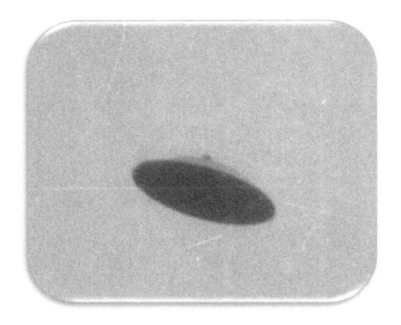

Enlarged View

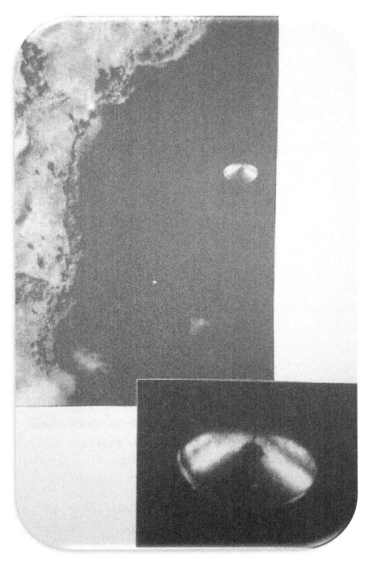

Brazil

CHAPTER 2
ROSWELL

*We shall postulate that on some distant planet, some 4 billion or so
years ago, there had evolved a form of higher creature who, like
ourselves, had discovered science and technology, developing them far
beyond anything we have accomplished.*

--Sir Francis Crick, discoverer of the DNA Helix.

Roswell happened in July 1947. The damning evidence
was a magnification of a telegram in the hands of a
Brigadier General Robert Ramey. In the infamous
photograph, you can see Major Jesse Marcel holding up a
weather balloon. Next to Marcel is Ramey, in whose
hand is... the telegram...Detailing that there were
victims of the wreck, that "FWAAF [Fort Worth Air
Force Base] acknowledges a disc is north - northwest of
Cordon", and to "assure that CIC/TEAM next set our
PR of weather balloons would fare better if... Weather
balloons where the cover story...that telegram is real."

Major Jesse Marcel graduated from radar school at
Langley, Virginia, and studied advance radar devices,
including all varieties of Rawin radar targets which
included the ML-307 reflector used on the Mogul device
(weather balloon). So Marcel was *very familiar* with all

types of weather balloons! How could the government use a weather balloon as a cover up story?

Evidence includes the following:

- RAAF (Roswell Air Force Base) Carried the nuclear bomb that was released in Hiroshima, Japan, during World War II,
  and was the only US nuclear facility in 1947. Nuclear experimentation may have attracted extraterrestrials to the site.

- On July 8th, the *Roswell Daily Record* ran the following headline: RAAF CAPTURED FLYING SAUCER ON RANCH IN ROSWELL REGION. (*Author - if it were a weather balloon, why use the words "capture" and "saucer"?*)

- Jesse Marcel confessed to Stanton Friedman (expert author who interviewed Marcel) that the disc was "was not from earth."

- Jesse Marcel took home pieces of a mystery metal that would straighten out after bending and was impenetrable to bullets.

- Jesse Marcel Jr. (his son) remembers seeing this mystery metal and remembering marking on it resembling hieroglyphics "of different geometric shapes, leaves and circles."

- John Kromschroeder, friend of W. O. "Pappy" Henderson, the pilot that flew the disk and debris to Wright Air Field in Ohio, said that he was given a piece of the metal by Henderson and that the metal was gray and resembled aluminum, but he was unable to cut it.

- Mac Brazel, owner of the field in which the crash took place, said, "I am sure what I found was not any weather observation balloon."

- "If the government were only transporting the wreckage of a Mogul balloon, it would not have needed an aircraft the size of a C-54, let alone guards to protect the over-the-counter materials that made up the balloons and instruments." (Marcel). (*Author-another smoking gun*)

- Glen Davis, a mortician working for Ballards Funeral Home in Roswell, said that he received a call from the Roswell Air Force Base requesting a few "small caskets" that could be "hermetically sealed." (*Author – for the miniature alien bodies*)

- Dan Dwyer, a firefighter in Roswell, said he saw the wreckage of a flying craft, two small dead bodies, and a "very small being about the size of a ten year old child."

- The impact area, about seventy-five miles northwest of Roswell, is at latitude thirty-three degrees north. This places it at 2012 nautical miles from the equator. When the latitude is multiplied by the universal mathematical constant pi (3.14), the result is 104 degrees. This is the precise longitude of the crash site. What are the odds of this particular crash site having these coordinates? Coincidence?

An Actual Mogul Balloon.

This Balloon caused so much panic?

Site of the Roswell Crash

The infamous photograph of General Ramey with Jesse Marcel holding up the fake weather balloon.

See the memo in Ramey's left hand? There's the smoking gun!

Enlargement of the Ramey telegram.

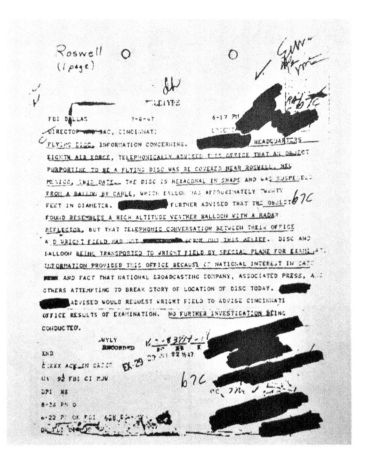

Roswell Memo dated 7/8/47 confirms:

- A disk was recovered near, Roswell, NM
- "Victims" of the wreck were forwarded to Ft. Worth, TX
- CIC teams would send PR of weather balloons

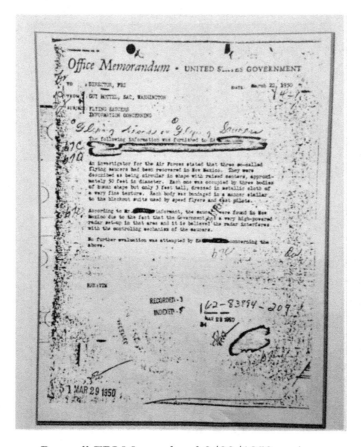

Roswell FBI Memo dated 3/22/1950 stating:

- 3 flying saucers had been recovered in New Mexico
- They were circular in shape with raised centers
- They were approximately 50 feet in diameter
- They were occupied by "three bodies of human shape but only 3 feet tall"

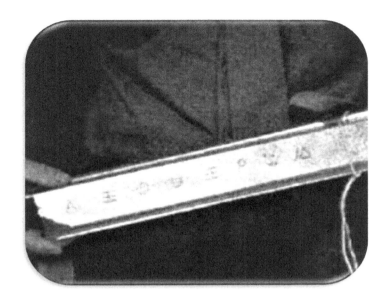

Actual I-Beam taken from Roswell crashed UFO

Actual Photo of alien hands controls from Roswell UFO

MJ-12

MJ-12 was a group of 12 specialists in various fields related to aerospace and military departments, established by President Truman in 1949, I attach a copy of a "TOP SECRET\MAJIC" manual (pgs. 142-158) addressing the steps necessary to collect and disseminate any disks and/or beings found, and the proper disinformation guides with which to address the public regarding it. A briefing document prepared in 1952 for President-Elect Dwight D. Eisenhower concerning *Operation* Majestic 12, reporting directly and only to the President of the United States, was to deal with the UFO issue at the highest level of government. The papers went on to say that "Since it is virtually certain that these craft do not originate in any country on earth, considerable speculation has centered around what their point of origin might be and how they they get here. Mars was and remains a possibility, although some scientists, most notably Dr. Menzel, consider it more likely that we are dealing with beings from another solar system entirely. Numerous examples of what appear to be a form of writing were found in the wreckage. Efforts to decipher these have remained largely unsuccessful" (Eisenhower Briefing document, November 18th, 1952).

- A secret operation was begun on July 7th, 1947 (two days after the crash) to recover the wreckage of a "disk shaped craft from a crash site approximately 75 miles northwest of Roswell Air Force Base."

- Four *humanoid beings* had apparently ejected from the craft and were found dead about two miles east of the wreckage site.

- News reporters were given the effective cover story that the object had been a misguided weather research balloon.

- The dead beings were termed "Extra-Terrestrial Biological Entities or EBEs."

- Efforts to decipher a form of writing found in the wreckage and to determine the method of propulsion were unsuccessful.

- Projects *Sign* and *Grudge* were implemented to determine the performance characteristics and purpose of this craft.

- Doctor Robert Sarbacher (a later member of MJ 12) told Wilbert Smith of the Canadian Department of Transport relative to radio-wave propagation and geomagnetism, when asked if flying saucers were real, "Yes they exist. They are classified two points higher than the H-Bomb." Smith then reported back to the Canadian government that "the entire matter is considered by the United States authorities to be of tremendous significance" (Rutlowski).

- FBI director J. Edgar Hoover wrote in a late-1947 document referring to the FBI's access to "discs" that "I would do it but before agreeing to it we must insist upon full access to discs recovered. For instance in the LA [Los Alamos] case the Army grabbed it and wouldn't let us have it for cursory examination."

- Albert Einstein and Robert Oppenheimer were called in to give their opinions, drafting a six-page paper titled, "Relationships with Inhabitants of Celestial Bodies."

- President John F. Kennedy was killed ten days after sending a memo to James Angleton on November 12, 1963, demanding to see all CIA UFO files. Kennedy informed Angleton (former Director of Counterintelligence) that he wanted to share this intelligence with Russia. (This information never got to JFK.)

The following were members of MJ-12:

1. Admiral Roscoe H. Hillenkoetter--first director of the CIA upon its formation in September 1947

2. Dr. Vannevar Bush—eminent American scientist who developed the Office of Scientific Research and Development, which led to the production of the first atomic bomb

3. Secretary James Forrestal--Secretary Of Defense in July 1947; resigned in March 1949, before "reportedly" committing suicide (Authors note-was he murdered?)

4. General Nathan Twinning--commander of the Air Force Material Command base at Wright-Patterson (allegedly where there are alien spacecraft *and* aliens kept to study); author of the infamous "Twinning Memo"

5. General Hoyt S. Vandenberg--Air Force Chief of Staff; ordered inception of *Project Sign*

6. Dr. Detlev Bronk—Chairman of the National Research Council and member of the Atomic Energy Council (allegedly the doctor who carried out the first alien autopsy)

7. Doctor Jerome Hunsaker--chairman of the Department of Mechanical and Aeronautical Engineering at MIT

8. Mr. Sidney Souers--executive secretary to the National Security Council and retired rear admiral

9. Mr. Gordon Gray--secretary of the army in 1949; special assistant to President Truman

10. Dr. Donald Menzel--director of the Harvard College Observatory and a respected astronomer

11. General Robert M. Montage--commander of Sandia Atomic Energy Commission facility in New Mexico
12. Doctor Lloyd Berkner--executive executive secretary of the Joint Research and Development Board in 1946

As you can see, MJ-12 consisted of experts in all areas of the military, sciences, and security. It's also of interest that the MJ 12 papers came to light after these members all were dead.

President John F Kennedy reportedly told a loadmaster for Air Force One after a UFO conference in Bonn, Germany, in 1963, "I'd like to tell the public about the alien situation, but my hands are tied."

In 1995, Linda Howe, a noted UFO researcher, received anonymously in the mail a copy of a secret MJ-12 document entitled, "SOM1-01 Majestic-12 Group Special Operations Manual--Extraterrestrial Entities and Technology, Recovery and Disposal. TOP/ SECRET MAJIC EYES ONLY." It was dated April 7th, 1954. It is essentially a Handbook for "Majestic-12 units" that are dispatched to crash sites and how to handle the debris and/or life-forms. In it, the extraterrestrial craft are referred to as "UFOBs" and the alien entities are referred to as "EBEs.". The document includes instructions such as, "Dead EBEs should be packed in ice" and sections on "How to deal with the Press" and "Press Blackout"--official denial, discrediting witnesses, and deceptive statements.

It describes 2 types of EBEs:

- Type I EBE--six fingers (same as once found in Roswell)
- Type II EBE—"the grays"—grey skin, big black eyes

Section 1.22 states, "Under no circumstance is the general public or the public press to learn of the existence of these entities. The official government policy is that such creatures do not exist, and that no agency of the federal government is now engaged in any study of extraterrestrials or their artifacts. Any deviation from this stated policy is absolutely forbidden" (Friedman).

The MJ-12 briefing document states "Aerial reconnaissance discovered that four small human-like beings had apparently ejected from the craft at some point before it exploded. These had fallen to the earth about two miles east of the wreckage site. All four were dead and badly decomposed due to action by predators and exposure to the elements during the approximately one week time period which had elapsed of their discovery. A special scientific team took charge of removing these bodies for study" (Friedman).

"The wreckage of the craft was also removed to several different locations. Civilian and military witnesses in the area were briefed, and news reporters were given the effective cover story that the object had been a misguided weather research balloon" (Friedman).

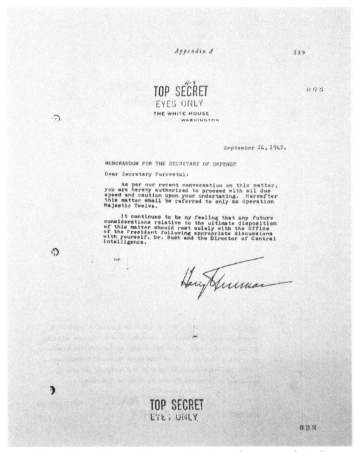

President Harry Truman's signed memo dated 9/24/1947, approximately three months after the Roswell crash, to the Director of the CIA, James Forrestal, initiating the formation of the secret group called "Operation Majestic Twelve" or MJ-12."

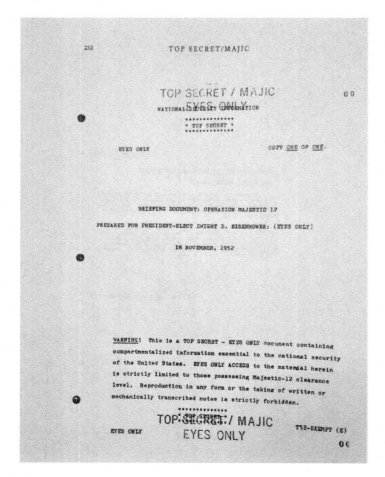

MJ-12 document prepared for the President Dwight D.
Eisenhower on 11/18/1952, tells the whole story.

TOP SECRET / MAJIC
EYES ONLY
* TOP SECRET *
****

002

EYES ONLY   COPY <u>ONE</u> OF <u>ONE</u>.

SUBJECT: OPERATION MAJESTIC-12 PRELIMINARY BRIEFING FOR
PRESIDENT-ELECT EISENHOWER.

DOCUMENT PREPARED 18 NOVEMBER, 1952.

BRIEFING OFFICER: ADM. ROSCOE H. HILLENKOETTER (MJ-1)

NOTE: This document has been prepared as a preliminary briefing
only. It should be regarded as introductory to a full operations
briefing intended to follow.

* * * * * *

OPERATION MAJESTIC-12 is a TOP SECRET Research and Development/
Intelligence operation responsible directly and only to the
President of the United States. Operations of the project are
carried out under control of the Majestic-12 (Majic-12) Group
which was established by special classified executive order of
President Truman on 24 September, 1947, upon recommendation by
Dr. Vannevar Bush and Secretary James Forrestal. (See Attachment
"A".) Members of the Majestic-12 Group were designated as follows:

Adm. Roscoe H. Hillenkoetter
Dr. Vannevar Bush
Secy. James V. Forrestal*
Gen. Nathan F. Twining
Gen. Hoyt S. Vandenberg
Dr. Detlev Bronk
Dr. Jerome Hunsaker
Mr. Sidney W. Souers
Mr. Gordon Gray
Dr. Donald Menzel
Gen. Robert M. Montague
Dr. Lloyd V. Berkner

The death of Secretary Forrestal on 22 May, 1949, created
a vacancy which remained unfilled until 01 August, 1950, upon
which date Gen. Walter B. Smith was designated as permanent
replacement.

****
* TOP SECRET *
TOP SECRET / MAJIC
EYES ONLY
EYES ONLY   T52-EXEMPT (E)

001

234

TOP SECRET / MAJIC
EYES ONLY
* TOP SECRET *
....

003

COPY ONE OF ONE.

On 24 June, 1947, a civilian pilot flying over the Cascade
Mountains in the State of Washington observed nine flying
disc-shaped aircraft traveling in formation at a high rate
of speed. Although this was not the first known sighting
of such objects, it was the first to gain widespread attention
in the public media. Hundreds of reports of sightings of
similar objects followed. Many of these came from highly
credible military and civilian sources. These reports resulted in independent efforts by several different elements
of the military to ascertain the nature and purpose of these
objects in the interests of national defense. A number of
witnesses were interviewed and there were several unsuccessful
attempts to utilize aircraft in efforts to pursue reported
discs in flight. Public reaction bordered on near hysteria
at times.

In spite of these efforts, little of substance was learned
about the objects until a local rancher reported that one
had crashed in a remote region of New Mexico located approximately seventy-five miles northwest of Roswell Army Air
Base (now Walker Field).

On 07 July, 1947, a secret operation was begun to assure
recovery of the wreckage of this object for scientific study.
During the course of this operation, aerial reconnaissance
discovered that four small human-like beings had apparently
ejected from the craft at some point before it exploded.
These had fallen to earth about two miles east of the wreckage
site. All four were dead and badly decomposed due to action
by predators and exposure to the elements during the approximately one week time period which had elapsed before their
discovery. A special scientific team took charge of removing
these bodies for study. (See Attachment "C".) The wreckage
of the craft was also removed to several different locations.
(See Attachment "B".) Civilian and military witnesses in
the area were debriefed, and news reporters were given the
effective cover story that the object had been a misguided
weather research balloon.

....
* TOP SECRET *
....
EYES ONLY TOP SECRET / MAJIC
EYES ONLY   T52-EXEMPT (E)

00:

236   TOP SECRET/MAJIC

TOP SECRET / MAJIC
EYES ONLY
* * * * * * * * * * * * *
* TOP SECRET *
* * * * * * * * * * * * *

EYES ONLY   COPY ONE OF ONE.

A need for as much additional information as possible about
these craft, their performance characteristics and their
purpose led to the undertaking known as U.S. Air Force Project
SIGN in December, 1947. In order to preserve security, liaison
between SIGN and Majestic-12 was limited to two individuals
within the Intelligence Division of Air Materiel Command whose
role was to pass along certain types of information through
channels. SIGN evolved into Project GRUDGE in December, 1948.
The operation is currently being conducted under the code name
BLUE BOOK, with liaison maintained through the Air Force officer
who is head of the project.

On 06 December, 1950, a second object, probably of similar
origin, impacted the earth at high speed in the El Indio -
Guerrero area of the Texas - Mexican border After following
a long trajectory through the atmosphere. By the time a
search team arrived, what remained of the object had been almost
totally incinerated. Such material as could be recovered was
transported to the A.E.C. facility at Sandia, New Mexico, for
study.

Implications for the National Security are of continuing importance in that the motives and ultimate intentions of these
visitors remain completely unknown. In addition, a significant
upsurge in the surveillance activity of these craft beginning
in May and continuing through the autumn of this year has caused
considerable concern that new developments may be imminent.
It is for these reasons, as well as the obvious international
and technological considerations and the ultimate need to
avoid a public panic at all costs, that the Majestic-12 Group
remains of the unanimous opinion that imposition of the
strictest security precautions should continue without inter-
ruption into the new administration. At the same time, contingency plan MJ-1949-04P/78 (Top Secret - Eyes Only) should
he held in continued readiness should the need to make a
public announcement present itself. (See Attachment "G".)

* * * * * * * * * * * * *
TOP SECRET/MAJIC
EYES ONLY
EYES ONLY   T52-EXEMPT (E)

e

TOP SECRET / MAJIC
A-4

EYES ONLY
•••••••••••••
• TOP SECRET •
•••••••••••••

004

EYES ONLY   COPY ONE OF ONE.

A covert analytical effort organized by Gen. Twining and
Dr. Bush acting on the direct orders of the President, resulted in a preliminary concensus (19 September, 1947) that
the disc was most likely a short range reconnaissance craft.
This conclusion was based for the most part on the craft's
size and the apparent lack of any identifiable provisioning.
(See Attachment "D".) A similar analysis of the four dead
occupants was arranged by Dr. Bronk. It was the tentative
conclusion of this group (30 November, 1947) that although
these creatures are human-like in appearance, the biological
and evolutionary processes responsible for their development
has apparently been quite different from those observed or
postulated in homo-sapiens. Dr. Bronk's team has suggested
the term "Extra-terrestrial Biological Entities", or "EBEs",
be adopted as the standard term of reference for these
creatures until such time as a more definitive designation
can be agreed upon.

Since it is virtually certain that these craft do not originate in any country on earth, considerable speculation has
centered around what their point of origin might be and how
they got here. Mars was and remains a possibility, although
some scientists, most notably Dr. Menzel, consider it more
likely that we are dealing with beings from another solar
system entirely.

Numerous examples of what appear to be a form of writing
were found in the wreckage. Efforts to decipher these have
remained largely unsuccessful. (See Attachment "E".)
Equally unsuccessful have been efforts to determine the
method of propulsion or the nature or method of transmission
of the power source involved. Research along these lines
has been complicated by the complete absence of identifiable
wings, propellers, jets, or other conventional methods of
propulsion and guidance, as well as a total lack of metallic
wiring, vacuum tubes, or similar recognizable electronic
components. (See Attachment "F".) It is assumed that the
propulsion unit was completely destroyed by the explosion
which caused the crash.

•••••••••••••
• TOP SECRET •
•••••••••••••

EYES ONLY TOP SECRET / MAJIC   T52-EXEMPT (E)

EYES ONLY

004

n-6
TOP SECRET / MAJIC
EYES ONLY
00(

```
**************
*  TOP SECRET  *
**************
```

EYES ONLY   COPY ONE OF ONE.

ENUMERATION OF ATTACHMENTS:

*ATTACHMENT "A".........Special Classified Executive
                        Order #092447. (TS/EO)

*ATTACHMENT "B".........Operation Majestic-12 Status
                        Report #1, Part A. 30 NOV '47.
                        (TS-MAJIC/EO)

*ATTACHMENT "C".........Operation Majestic-12 Status
                        Report #1, Part B. 30 NOV '47.
                        (TS-MAJIC/EO)

*ATTACHMENT "D".........Operation Majestic-12 Preliminary
                        Analytical Report. 19 SEP '47.
                        (TS-MAJIC/EO)

*ATTACHMENT "E".........Operation Majestic-12 Blue Team
                        Report #5. 30 JUN '52.
                        (TS-MAJIC/EO)

*ATTACHMENT "F".........Operation Majestic-12 Status
                        Report #2. 31 JAN '48.
                        (TS-MAJIC/EO)

*ATTACHMENT "G".........Operation Majestic-12 Contingency
                        Plan MJ-1949-04P/78: 31 JAN '49.
                        (TS-MAJIC/EO)

*ATTACHMENT "H".........Operation Majestic-12, Maps and
                        Photographs Folio (Extractions).
                        (TS-MAJIC/EO)

```
*************
*  TOP SECRET  *
```
TOP SECRET / MAJIC
EYES ONLY
EYES ONLY   T52-EXEMPT (E)

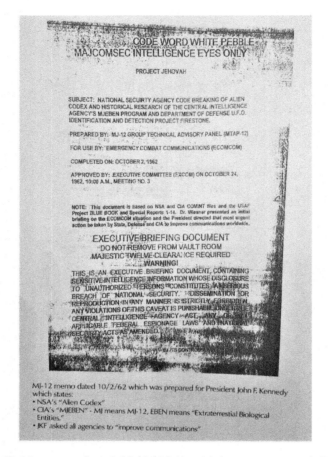

MJ-12 memo dated 10/2/62 which was prepared for President John F. Kennedy which states:
• NSA's "Alien Codex"
• CIA's "MJEBEN" - MJ means MJ-12, EBEN means "Extraterrestial Biological Entities."
• JKF asked all agencies to "improve communications"

MJ-12 memo dated 10/2/1962 which was prepared for President John F. Kennedy which states:

- NSA's "Alien Codex"
- CIA's "MJEBEN" – MJ means MJ-12, EBEN means "Extraterrestial Biological Entites"
- JFK asked all agencies to "improve communications"

CHAPTER 3
EISENHOWER BRIEFING DOCUMENT

Excerpts from Operation Majestic 12, Briefing Document, prepared for Dwight D. Eisenhower-- November 18th, 1952

"Dr. Bronk's [member of MJ-12] team has suggested the term 'Extra-terrestrial Biological Entities', or EBES."

"A need for as much additional information as possible about these craft, their performance characteristics and their purpose led to the undertaking known as U.S. Air Force Project SIGN in December, 1947."

"SIGN evolved into Project GRUDGE in December, 1948. The operation is currently being conducted under the code name BLUE BOOK."

"On 6, December, 1950, a second object, probably of similar origin, impacted the Earth at high speed in the El Indio Guerrero area of the Texas-Mexican border after following a long trajectory through the atmosphere."

"It is for these reasons, as well as the obvious international and technological consideration and the ultimate need to avoid a public panic at all costs."

Excerpts from SOM1-01 (Special Operations Manual); Extraterrestrial Entities and Technology, Recovery and Disposal: April, 1954

- "These objects and their builders pose a direct threat to the security of the United States, despite the uncertainty as to their ultimate motives in coming here."
- "Certainly the technology possessed why these beings far surpasses anything known to modern science, yet their presence here seems to be benign, and they seem to be avoiding contact with our species, at least for the present."
- "Several dead entities have been recovered along with a substantial amount of wreckage and devices from downed craft , all of which are now under study at various locations."
- "No attempt has been made by extraterrestrial entities either to contact authorities or to recover their dead counterparts of the downed craft, even though one of their crashes was the result of direct military action."
- "Metallurgical analysis of the bulk of the debris recovered indicates that the samples are not composed of any materials currently known to terrestrial science."
- "Much of the material, having the appearance of aluminum foil or aluminum magnesium sheeting , displays none of the characteristics of either metal,

resembling instead some kind of unknown plastic-like material."

- "Several samples were engraved or embossed with marks and patterns. These patterns were not readily identifiable and attempts to decipher their meaning has been largely unsuccessful."

"Witnesses will be discouraged from talking about what they have seen, and intimidation may be necessary to ensure their cooperation."

232

TOP SECRET/MAJIC

TOP SECRET / MAJIC
EYES ONLY   0 0
NATIONAL SECURITY INFORMATION
••••••••••••••
• TOP SECRET •
••••••••••••••

EYES ONLY   COPY ONE OF ONE.

BRIEFING DOCUMENT: OPERATION MAJESTIC 12

PREPARED FOR PRESIDENT-ELECT DWIGHT D. EISENHOWER: (EYES ONLY)

18 NOVEMBER, 1952

WARNING! This is a TOP SECRET - EYES ONLY document containing
compartmentalized information essential to the national security
of the United States. EYES ONLY ACCESS to the material herein
is strictly limited to those possessing Majestic-12 clearance
level. Reproduction in any form or the taking of written or
mechanically transcribed notes is strictly forbidden.

••••••••••••••
• TOP SECRET •
TOP SECRET / MAJIC
EYES ONLY   T52-EXEMPT (E)
EYES ONLY
                                    0 0

This unclassified copy is for research purposes. Rev 3; 2/28/95; Helvetica headers, Times text

RESTRICTED

SOM1-01

TO 12D1—3—11—1

MAJESTIC—12 GROUP SPECIAL OPERATIONS MANUAL

EXTRATERRESTRIAL ENTITIES AND TECHNOLOGY, RECOVERY AND DISPOSAL

TOP SECRET/MAJIC EYES ONLY

WARNING! This is a TOP SECRET—MAJIC EYES ONLY document containing compartmentalized information essential to the national security of the United States. EYES ONLY ACCESS to the material herein is strictly limited to personnel possessing MAJIC—12 CLEARANCE LEVEL. Examination or use by unauthorized personnel is strictly forbidden and is punishable by federal law.

MAJESTIC—12 GROUP • *APRIL 1954*

MJ—12 4838BMAN 270435¹-54-1

CHAPTER 4
THE
EXTRATERRESTRIALS

Through FOIA (Freedom of Information Act), the following information concerning extraterrestrials has been uncovered:

There are fifty-seven species of aliens, all humanoid that have been found in sixteen crashes across the globe and may be distant cousins of the Anunnaki. (discussed in Chapter 12)

The first EBE "spoke" in tonal variations.

According to Len Kasten in *The Secret History of Extraterrestrials,* the first ET captured after the Roswell crash came into contact with his home planet—Serpo. Serpo is in the Zeta Reticuli system, about forty light years from earth. After the death of the EBE, Los Alamos, New Mexico, scientists used equipment found at the Corona, New Mexico, crash in order to communicate with the planet. According to Kasten , they received the first reply in December 1952! Incredible information!

ETs have asked us to "beat our swords into plowshares" (an exact comment that Ronald Reagan made at a speech regarding Russia) meaning to give up our nuclear weapons.

ETs are apparently brought to and stored at Wright-Patterson Air Force Base in Dayton, Ohio.

Captured ETs had an octagon-shape crystal which, when positioned a certain way, showed pictures of their home planet.

EBES

While EBE #1 was alive, there was apparently a failure to

communicate between EBE #1 and our military, They failed to ask EBE #1 about the CR [crystal rectangle] device or any other items abroad EBE's spacecraft except

one. It was a small (12" x 9" x 2") device. It contained several small holes, two 4" antennas and two black inlayed

"chips". (Author's note -computer chips). The communi-

cation device also had a series of lights that would alternate

from left to right when an incoming signal was received. The lights would alternate from right to left when an out-

going signal was sent. What was interesting was how this device was powered. It was later learned that the commu-

nication device was connected to the CR by a small glass tube (Author's note – Fiber optics?). There were no wires

within the glass tube. To the best of Mr. R's [Author's note

-Richard Helms, former Director of the C.I.A. from 1966-1973] knowledge, they still haven't figured out how the power goes from the CR to the communication device.

Although we could not operate the device, we used it to manufacture [another] device used to communicate with the Ebens. Mr. R states that the device was classified until

the late 70's now the military uses it (Collins).

The first EBE came from a planet called Sieu (our planet identified as Serpo) in the star group of Zeta Reticuli. This star group is 220 trillion miles from earth. The planet is thousands of years ahead of us, and the Ebens have been navigating the solar system thousands of years as well. From this information, we created what's called *The Yellow Book*, our version of the history of the universe. Another book called *The Red Book* is a summary of UFO/EBE investigation from 1947 on. It is updated every five years. Some interesting facts about the Ebens physiology--Their hearts and lungs are one organ, connected. They enjoy music, especially Tibetan music and chants. They can travel from Serpo to earth in approximately ninety days.

Taken from an interview with an anonymous whistle-blower: "ET's had large heads and were around 4 feet tall. They have small noses and mouths with no ears or hair. The hand has 4 fingers on it with one finger twice as long as either outside finger. Brain capacity is 1800 cc versus 1300 cc for the average human. The skin is grey or ashen an under the microscope appears mesh-like.

There was a colorless liquid in the body without red cells, no lymphocytes, no hemoglobin. There was no digestive system, intestine, alimentary canal, or rectal area in the ET autopsy" (Komarek).

TOP SECRET
CENTRAL INTELLIGENCE AGENCY
WASHINGTON, D.C. 20505

HANDLE ON STRICT
NEED-TO-KNOW BASIS

May 11, 1980

HCDRS

RA-111

REF: Memo 5-7A

The Texas incident will present a problem for us. Must
determine more info in order to develop dis-info plan.
Why do these things always happen when MJ12 is gone. Luck
is never on our side. Let R-J know something is up and
that we will need their support to get through this one.

Don't make anymore phone calls on KY1 until the new code is
changed. Too many of the wrong people might be listening.
I don't trust B-2A. Could be leak there.

One more thing, EBE-2 has expressed a desire to visit ocean.
I don't know what the hell to do on that one. If we don't
let him, he'll just disappear. Can't allow that to happen
again. MJ6 is working on that. TREAD open/or will be.

CIA memo on EBEs

AREA 51

"The Use Of Deadly Force Is Authorized." This is what a sign reads prohibiting anyone entering Area 51 via the public highway in New Mexico.

Obviously, the Air Force is hiding something. Now, you may say, "Of course, it's obvious. They don't want any military secrets of new, high tech planes to leak." And that's true—the stealth bomber was designed and tested here. But there's more to it. It is a perfect hiding place for alien craft. Think *Close Encounters Of The Third Kind*, which I believe leaks (prepares us for) the truth. One employee of area 51, Bob Lazar is a very controversial figure. He worked for the US government at Area 51 in 1982 in advanced propulsion. At Dreamland / the Ranch / Skunk Works (various other names for Area 51), he claims he was reverse-engineering a power plant from a UFO--a matter/antimatter reactor. He said the reactor was very similar to our nuclear reactors, but it was "the size of a soccer ball." He said that this reactor "produced waves nullifying gravity through the use of a super heavy substance he named Element 115 after its theoretical atomic number." Laser claims that at area 51, the United States is studying gravity drives, time travel, and a neutron beam weapon. He claims that the aliens refer to the earth as "SOL 3," (Author's note--third planet from the sun) that mankind is the result of sixty-five genetic "corrections" made over thousands of years and at least one group of small ETs, referred to as "the kids," come from a planet called Zeta Reticuli. Aliens consider humans as "containers"--containers of what? Lazar continues, "Maybe containers of souls. But we're containers, and that's how we are mentioned in the documents; that religion was specifically created so we have rules and regulations for the sole purpose of not damaging the containers. (Marrs, *Alien Agenda*).

UFOs have been seen above bodies of water. Perhaps they are extracting hydrogen from the H2O and using it as fuel?

Their movement has been associated with a local distortion of space time, using the force ahead of it to thrusted forward like a surfboard along the waves, faster than the speed of light.

More on propulsion in a later chapter.

CHAPTER 5
THE MOON

"The placement of five nuclear-powered scientific stations on the lunar surface" (Marrs *Alien Agenda*.)

Rocks discovered on the moon have an age of 4.5 billion years. One moonrock was dated at 5.3 billion (Harvard Astronomy Journal, *Sky And Telescope*). The oldest earth rocks are 3.5 billion years.

Does water exist on the moon? While there is none on the surface, a "wind" of water was detected on March 7th, 1971, by instruments left behind by Apollo missions, which had sent a signal to earth. This was proclaimed "one of the most exciting discoveries yet" by Rice University Physicists Dr. John Freeman and Dr. H. Ken Hills. However, the most interesting part is that the "wind" didn't originate from outside the moon but rather *inside* it! Apparently, the "wind" was released during a moonquake from deep within the core. Additional evidence of water on the moon can be demonstrated by moon rocks containing small pieces of rusted iron, indicating oxidation (oxygen and free hydrogen molecules). Moon rocks were also found to be slightly magnetic, even though there is no magnetic field on the moon itself.

PYRAMIDS ON THE MOON

In November 1966, *Lunar Orbiter 2* took photos of what appeared to be pyramids. NASA, of course, denied that it photographed anything unusual. Russian space engineer Alexander Abramov stated, "The distribution of these lunar objects is similar to the plan of the Egyptian pyramids." Was JFK's race to the moon less about beating the Russians to the punch and more about verifying and documenting these pyramids?

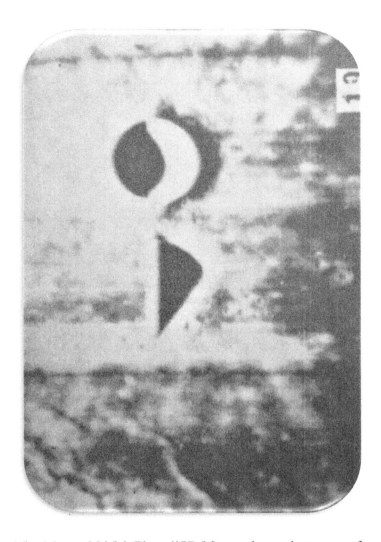

The Moon. NASA Plate #57. Moon photo shot west of the crater Aristarchus. These are two artificially made circular and triangular installations. The triangular installation is purportedly 25 miles long!

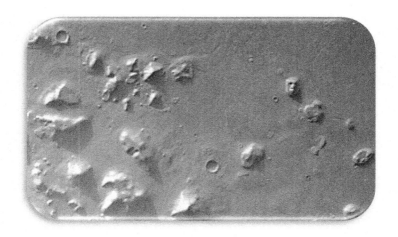

Viking Image of the Cydonia region on Mars. The "face" is on the back right.

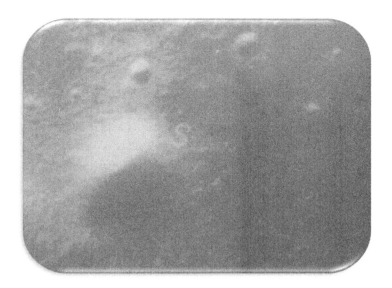

Apollo 14 photo No. 14-80-10439. See the serpent like "S" symbol. Serpents come up in later chapters from the Bible to representation of the DNA helix.

SPACECRAFT ON THE MOON

During *Apollo* 11's infamous landing in the Sea of
Tranquility, Otto Binder , an ex-NASA space program
member, claims that ham radio enthusiasts who listened
to the landing overheard one of the astronauts exclaim,
"These babies are huge, sir ... enormous ... Oh God,
you wouldn't believe it! I'm telling you, there are other
spacecraft out there ... lined up on the far side of the
crater edge ... They're on the moon watching us."

The term "bogey," commonly meaning a UFO, has been
used on film footage on many spaceflights to the moon.
Even on YouTube TM, you can hear the astronauts
saying, "I can see a bogey to the north-northeast" and so
on. Bogey's refer to alien spaceships. They are
photographed with white lights outside of our rockets'
windows. There is a very famous YouTube video of a
piece of a ship (a tether) falling off into orbit and *hundreds*
of alien craft of all shapes and sizes flying past the piece
in space--some fast, some slow, some disc-shaped, some
circular, and one in particular that appears to be spinning
very rapidly. Also, these ships are huge, based upon the
size of the falling tether. How could this be a hoax! It is
an actual NASA footage of a piece of man-made material
falling through space. There are multiple dialogues going
on throughout the process. At one point, one person
asks the other if they can see the "debris" moving by the
falling tether, and the other party says "yes." This was no
debris. Take a look and see for yourself.

According to author Timothy Good, astronaut Neil Armstrong is purported to have said at a NASA symposium, "It was incredible ... The fact is, we were warned off. There was never any question then of a space station or a moon city ... I can't go into details, except to say that their ships were far superior to ours in size and technology-- boy, were they big ... and menacing."

Maurice Chatelain, former chief of NASA communication systems, said in *Nexus* magazine and April 1995, "Neil Armstrong saw two UFO's on the rim of the crater," and "All Apollo and Gemini flights were followed ... by space vehicles of extraterrestrial origin ... The astronauts informed Mission Control, who then ordered absolute radio silence."

What did we find on the moon, and why haven't we returned after fifty years!? Is it what we were told to believe--a dark, dead planet without life? Or is it an ET way station that scared the hell out of us? I believe the NASA photographs that I have seen have been airbrushed. NASA has deleted all signs of spacecraft or alien technology built on the moon. If you look closely at some of the photographs, you can see craters photoshopped into the scene. Why were these craters inserted? There were also several photographs showing light emanating from different areas of the moon that was not "our" light.

Could it be that ETs had been colonizing and mining on the moon? We know that there is a super abundance of titanium on the moon. We also know that nitinol, a memory metal that we humans developed with the help of a crashed spaceship, is made with nickel and titanium. So maybe the ETs had been mining the moon for titanium.

The Original Sphinx. Compare its face and headdress to that of the "Face" on Mars.

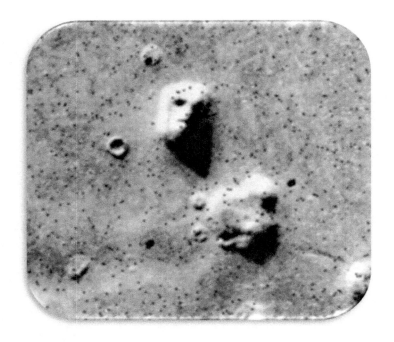

The infamous and real, "Face" on Mars. It looks like Egyptian because it was built by the Anunnaki teacher of the Egyptians.

Plate 102. NASA Photo No. AS-12-497319. White glowing oval object hovering over Apollo 12 Astronaut.

CHAPTER 6
THE PYRAMIDS

Nobody sets out to build 50,000 plus pyramids and mounds around the planet just for the hell of it ... These primitive people were hunter-gatherers who spent their waking hours running down there next meals.

Carl P. Munck

Why three pyramids in Giza? There was *one* pharaoh who ruled Egypt at a time. Why three--two large and one smaller? Why was the smallest offset to the East? Were all three built as a unified plan?

I believe that both the pyramids and the Sphinx are well over 5000 years old! Perhaps even 12,000. I also believe it was impossible for humans to have built these historic monuments. The Great Pyramid of Giza is 481 feet high. It was constructed with over 2.5 million massive stone blocks, each stone weighing 5,200 pounds for an estimated total weight of over 6 million tons (the weight of 60 aircraft carriers). The stones used to build the pyramids originated in quarries in Tura on the *other side* of the Nile, more than *600 miles away*! The white limestone blocks are held together by an amazing form of concrete that is not reproducible today. The foundation is designed to flex with hot and cold expansion and contraction and to with stand earthquakes. The concrete used was made from a strange, high technology composite that had tiny nanospheres in it. It is impossible to imagine the ancient Egyptians using only animal power and primitive tools building such a structure. Even if they use massive manpower to lift the stones, they could never have been lifted in exact proportions per tier as needed to create the pyramid shape.

If the ancient Egyptians did in fact use a roller system (trees) to haul the massive stones, then where did the trees come from? There are no trees in the vicinity. This is barren desert. The weathering of the Sphinx alone shows evidence of rainwater erosion, which has been scarce to none in Egypt since the last Ice Age. Robert Bauval, whom I met recently and had breakfast with, a master Egyptologist, tells us that the Giza pyramid is laid out to reflect the sky map (based on his computer evidence of the sky signature at the time) of the age of Leo, and this Sphinx (representing the lion) points directly to Leo at sunrise on the vernal equinox (the beginning of the ancient new year). This would make the pyramids age at 10,500 BC. This is when the pyramids would have aligned precisely with the Orion constellation at this time period.

You will read in your schoolbooks--and current Egyptologist will tell you on television--that all three pyramids were built during one century in 2500 BC using only the following:

• Stone

- Copper Tools
- Ropes (from plant fibers)
- Wood sledges

No wheeled vehicles, pulleys, or animals were used. So let's do the math ... Six million tons of stones were cut, hauled, shaped, and positioned in only one hundred years? That's five thousand tons of stones per month, or seven every hour. How can this be possible? It's not, because it's *impossible!*

It is also a farce to believe that what we are told in the classroom--that the pyramids were built as tombs of the dead pharaoh's--can be true. There is no evidence of this. None. Upon excavating inside the pyramids, there were *no* dead pharaohs and only *one* supposed tomb, which was way too small to house a dead body. I believe that its purpose was to house an instrument of electrical generation that could "start-up" the pyramids electrical properties. And even if grave robbers came into the pyramid to steal any/all artifacts there, they would not have touched any dead bodies. Where are all these tombs? There aren't any, and there never were. The pyramids had *nothing to do* with bearing the dead. The pyramids had everything to do with the needs of our makers.

PYRAMID PLACEMENT

While the Sphinx reflects the age of Leo, the positions of the 3 Giza pyramids match *precisely* the 3 stars of Orion's Belt, and the spatial relation of the pyramids to the Nile River is *exactly* the same as the relation of Orion to the Milky Way at the time. Coincidence? Thus the Sphinx and the pyramids of the Giza plateau represent on the ground what appeared in the sky at sunrise on the vernal equinox! While the Giza pyramids reflect the Orion constellation, the Nile River reflects the Milky Way. The Red pyramid location symbolizes the Sun, the bent pyramid location symbolizes the earth, the position of the Sirius Star System corresponds to the location of Heliopolis on the ground (McDowell).

So are we to surmise that the ETs who built the pyramids were from either Orion or Sirius? Why these two stars? We will go on to read that our forefathers were from planet Nibiru, not Orion or Sirius. So why honor these two stars?

The whole concept of the pyramids being used solely as tombs for dead pharaohs is a myth! There is not a single shred of evidence in writing or physical evidence that proves this myth. Robert Bauval as conclusively demonstrated to the world that one of the primary functions of the Giza complex was to serve as a ground map of the stars that appear overhead in direct proportion—the constellation of Orion. Bauval used computer renderings to figure out how the sky would look to those humans on the ground. Trying year by year, he could see the stars shift . When the computer reached *exactly* 10,500 BC , the alignment of these three pyramids of Giza *perfectly aligned with* the three stars in Orion's belt directly above. The southern shaft of the Kings chamber pointed toward the belt of Orion (which is associated with Osiris), and the Queens chamber pointed towards Sirius (associated with Isis). These alignments were not accidental. Obviously, whoever built the pyramids was mimicking those three stars. Why *those* three stars?

View of the three pyramids from above.

A view of Orion's Belt.

Can the striking similarity of the placement of the two be
mere coincidence?

PYRAMID PURPOSE

Don't you think that if the pyramids were built by human slaves under the direction of Egyptian masters--and their purpose was to glorify their dead gods--that *something* would have been written to identify the master maker? Well, there's *nothing*. No symbols, no inscriptions, no nothing. That's because it wasn't built by human slaves for their human masters!

Now here's where it gets really interesting. Christopher Dunn, an engineer by trade and Egyptologist, has analyzed the construction of the Great Pyramid from an engineering perspective and concluded that it was a *power generation plant* in which hydrogen generated in the Queen's chamber was excited by resonant *sound waves* in the King's chamber to emit microwave energy out it's southern shaft! We will read about the effects of sound later (levitation), how the use of sound--drums, trumpets--can stimulate movement of matter. Sound acts as a spark plug, using the "ether" in the air to its benefit and ultimate objective.

Zecharia Sitchin (discussed in Chapter 12) believed that Sumerian records told of the equipment in the Great Pyramid that produced a pulsating beam that guided aerial vehicles between orbiting space stations and the earth.

Looking at the top of the Great Pyramid, one can easily see that the apex has been removed and replaced. What happened to it? As we will read about our forefathers, the Enki and the Enlil clans were antagonistic. The Giza complex was in the territory of Enki while Jerusalem was the territory of the competing Enlil. According to Sumerian texts, at the end of a war between the two Anunnaki clans in 8670 BC, Ninurta (an Enlil leader) supposedly entered the Great Pyramid and removed and/or destroyed some system components. Included in the removal is whatever was mounted at the now *missing top* of the Great Pyramid. Ninurta complained of deadly radiation that may have emanated from a microwave antenna at the apex of the pyramid. All of this is written in the Sumerian texts.

Advanced machining was used to cut and form stone blocks in ancient Egypt. You can see from the photos that even today it is almost impossible to recreate the precision cuts in such deep and heavy limestone. It is impossible to create those lines from copper tools. No matter how careful the craftsman, these lines are impossible to have been made by human hands. They are the work, I believe, of Anunnaki lasers, using the value of pi and geometry as their architectural plan.

The Jewish Star of David consisting of two triangles overlaid, one pointing down and the other pointing up, may be representative of two things—the Great Pyramid communicating upward with "heaven" and another pyramid on Nibiru (the twelfth planet and home of the Anunnaki) communicating downward to earth. Another idea is that it represents copulation/reproduction/life-- the phallus and the uterus. The notion of "heaven" being upward in the sky (as we have been taught to believe) may actually be representative of Nibiru, our mother planet. I believe that that's *exactly* what heaven is--the planet Nibiru, a funny-sounding planet, well documented on Sumerian clay tablets *as heaven from which the gods came* (more on this in later chapters).

Were the pyramids a repository for nuclear fuel and weapons? Radiation levels exist in Giza today. The pyramids would have protected humans from radiation on the outside ...

"This evidence stands in plain sight in Egypt, whose ancient monolithic structures *are not tombs for pharaohs but the very real remains of a radio transmitter for messages to their home planet.*" Dr. Konstantin Meyl, a German professor of engineering, believes that the pyramids were a scalar radio technology based on electromagnetics. They were sophisticated radio transmitters and receivers to their home planets. The tablets inside (presumably the Ten Commandments), contained the power source necessary to activate the receiver-transmitter. It's also interesting that the original Sumerian word interpreted by Hebrew scribes as "salt" also meant "vapor". Vapor may have been the byproduct of these transmissions.

Obelisks today, such as the Washington monument and hundreds of others worldwide, symbolize the rocket ships of old ... tombstones with round tops symbolize the circular nose of space ships in the past ...

Why are these pyramids in Sumer, Egypt, India, Mexico, Peru, and China? How did both the Sumerians and Mayans acquire advanced astronomical and mathematical knowledge? Why did these civilizations also abandon the hunter-gatherer (meat-eating) way of life in favor of agriculture? How can this be, given that there was no communication between the two civilizations? Or was there? I believe there was …

The Great Pyramid of Shensi, located forty miles West of Xian in China, is not written about in our schoolbooks. It is comprised of thirteen levels and is missing the capstone (top), just like Giza. It is also--like Giza--aligned precisely according to the compass points. Unfortunately, the Chinese government has banned further exploration of this pyramid.

Interestingly as well, our one-dollar bill has a missing cap stone on the back with an "eye" in it … coincidence?

Some amazing facts about the pyramids include the following:

- The Great Pyramid is oriented exactly according to the four points of the compass.
- The Great Pyramid is located at the center of the earth's land mass.
- The three pyramids of Giza form a Pythagorean triangle.
- The Great Pyramid is a giant sundial. Its shadows reflect the seasons and the length of the year.
- The base of the pyramid equals 365 yards . Coincidence?
- The distance between the Great Pyramid and the North Pole is equal to the distance between the Great Pyramid and the center of the earth.

- The surface area of the four pyramid sides combined is equal to the pyramid height squared.
- Experts have found machining marks on carved granite inside the great pyramids, indicating the use of a powered rotating grinding wheel.

(Von Daniken)

WHO BUILT THE PYRAMIDS?

Whoever built the pyramids knew the precise circumference of the planet , the length of year, the length of the earth's orbit around the sun, the density of the planet, the cycle of the equinoxes, acceleration of gravity … and so on, and so on, and so on. How could they possibly *know this* when the earth was considered flat until the seventeenth century? Please ask yourself this question. If you really truly don't believe me, how can this be possible? If it *was* possible, then why aren't the Egyptians and not Sir Isaac Newton known for these breakthroughs in science? Why isn't there a builder (a specific, flesh and blood human) whose name is associated with the pyramids? There is none.

The pyramids were built by advanced, highly intelligent beings with full knowledge of their environment . The Egyptians inherited the pyramids; they didn't make them. It was in the process of development like an automobile which refines overtime It just *was*.

What is the logical explanation for the similarity of pyramids in Egypt, South America, and China? How can it be that they are so similar if ships never sailed across the Atlantic Ocean, and communication never occurred between the races? Why build a mound in the shape of a pyramid? Why not square?

Another theory suggests the pyramids of the earth are located at specific tetrahedron points on earth--that their exact positions are at points that amplify or harness electromagnetic energies of the planet earth . I believe in this theory. The pyramids weren't just made in random locations.

"There is also belief that a great Hall of Records exist under the paws of the Sphinx. According to Edgar Cayce, part of these priceless records holds the history of Atlantis, it's people and ultimate destruction. Recently, explorer Robert Schoch, while exploring the Sphinx with underground radar, had 'discovered' clear evidence of a cavity or chamber under the left paw of the Sphinx" (Redfern).

The Maya appear to have achieved what the Egyptians accomplished with the Great Pyramid: They created a massive geodesic marker in what is the 'naval' or geographic center, of the land mass of North and South America—a feat requiring a sophisticated command of accurate astronomical, mathematical, and geographical knowledge" (Hart).

The following translation is taken directly from Sumerian tablets: "To honor the builder, they agreed to build a monument nearby, with the face of the builder and the body of a lion, which symbolized the age during which it was built [Leo] ... Let us put beside the twin peaks a monument create, the Age of the Lion it announces ... The face of Ningishzidda [See Anunnaki chart], the peak's designer, let its face be ... toward the place of celestial chariots [spaceships] gaze" (Hart).

Many people think that the Sphinx has eroded overtime due to wind, but it's not true. The erosion is a direct result of the water and rain that occurred possibly 10,000 years ago when Giza had been a tropical, lush climate. This is why the Sphinx dates to 10,000 years ago and not 2,500 years ago.

Chichen Itza offers a spectacular occurrence every March and September 21st. "The Castle" is the predominant pyramid. It is dedicated to the feathered serpent called Kukulcan. On each of these days at the precise time of the equinox, the pyramid throws off a series of triangular shadows on the side of the main staircase. The shadow simulate the body of an undulating snake connecting to the serpent head built into the side of the staircase. In March the serpent descends, and in September the serpent ascends the staircase of the pyramid. This phenomenon last for over 3 hours. Could this be random coincidence? Who, without complete mastery of physics and astronomy could complete such a design? Even today, to complete such a stone "machine" would be next to impossible.

It is a sign of our makers--makers being *plural*. The pyramids exist as a physical sign that we were born of highly intelligent beings from another planet . Adam and Eve were born from this race. They were the first humans produced by the Anunnaki race.

WHY PI?

The number of pi 3.14(5)--has come up in a number of pyramids and monuments alike. The surfaces of these structures mostly utilized pi. The Giza pyramid and the Tihuanaco both used this method. It seems absurd to me to believe that resident Egyptians or South Americans used pi randomly when they decided to build a pyramid!

One of the largest, yet rarely discussed pyramids in the
world—The Great Pyramid of Shensi--China

CHAPTER 7
PROOF OF VITRIFICATION

Sodom and Gomorrah in the Sinai Peninsula in Israel demonstrate proof of the explosion of a nuclear device. An analysis of rock samples tested at the site contained only the isotope U-235 and none of the far-more-prevalent isotope U-238. The scientists (who have requested anonymity) initially concluded that the purity of production of the U-235 has not yet been achieved by modern human. A second observation was that the limestone surface of these rocks was impacted by a very high heat that discolored them and that they were unquestionably impacted by a nearby nuclear heat source for a short duration. There is no proof at this heat source was generated by an asteroid impact, because no shock grains were found, and there is no evidence of a volcanic eruption, because they are sedimentary (not igneous) rocks.

<u>Sodom and Gomorrah:</u>

Sodom and Gomorrah was surely a nuclear blast, turning those that looked at it in its vicinity to a "pillar of salt" and making their "hair fall from their head" (coincidence?). Could the radioactive fallout from this nuclear war have been the reason for Abrahams migration westward towards Egypt? It seems plausible to me. And how else could someone turn into a pillar of salt (figuratively) other than having been near a nuclear blast? Why would the Sumerian tablets talk about hair falling "from their heads?" We know today that this is the direct result of human interaction with radiation and its effects on our bodies. Why else would these words be chosen to be written?

Yahweh is also written on Sumerian tablets as being the ET who destroyed Sodom and Gomorrah.

I am an avid reader of books on alternative sciences, UFO's, and evolution. I believe firmly that the Sumerian tablets that have been translated brilliantly by Zecharia Sitchin and other scholars tell a **true story.** So do the Ramayana, the Mahabharata, and other Hindu texts.

The Hindu texts, which are very detailed, speak of a highly technical extraterrestrial race going to war with each other on earth! Proof of this war is vitrification (evidence of nuclear blasts) on several continents. Why the extraterrestrials went to war is not important in my work. The fact that we have *physical evidence* of this *is*. I have attached photos of vitrification in areas that have not had nuclear war in the twentieth century.

Here is an excerpt from the Mahabharata, written over eight thousand years ago:

Gurkha flying in his swift and powerful Vimana [UFO] hurled against the three cities of the Vrishis and Andhakas a single projectile charged with all the power of the Universe. An incandescent column rose up smoking and fire, as brilliant as ten thousand suns, rose in all its splendor. It was the unknown weapon, the Iron Thunderbolt, a gigantic messenger of death which reduced to ashes the entire race of the Vrishis and Andhakas [Sodom and Gomorrah?], the corpses were so burned as to be unrecognizable. Hair and nails fell out; pottery broke without apparent cause. (Komarek)

Another example of vitrification lies in the sands of Egypt. In 1932, Patrick Clayton, a surveyor for the Egyptian Geological Survey, was driving in the southwestern corner of Egypt when he heard his tires crunch on something that wasn't sand! It turned out to be large pieces of clear, yellow green-grass. It wasn't man made in the form of bottles. It was ultrapure 98% silica. This glass is a by-product of a nuclear explosion.

Doctor Robert Oppenheimer, chief scientist of the Manhattan Project, quoted from ancient Sanskrit—the Bhagavad Gita--in an interview after he watched the first atomic test: "Now I am become Death, the Destroyer of Worlds." Seven years later, in an interview at Rochester University in New York about whether the Alamogordo nuclear test was the *first* atomic bomb ever to be detonated, his response was this: "Ancient cities in brick and stone walls have literally been vitrified, that is, fused together, can be found in India, Ireland, Scotland, France, Turkey, and other places. There is no logical explanation for the vitrification of the stone forts and cities, except for an *atomic blast*." (Komarek)

In an article entitled, "*Ancient City Found in India, Irradiated From Atomic Blast,*" it states that a heavy layer of atomic ash has been found in Rajasthan, India, Where a housing development was scheduled to be built. There had been a high rate of cancer and birth defects associated with this area, and, while inspectors from the Indian government sealed off the area, they found and unearthed an ancient city dating back eight thousand to twelve thousand years. This city show destroyed buildings and the death of thousands of people. It is estimated that the nuclear blast was akin to the Hiroshima blast of 1945. (Komarek).

Vitrification. Proof of a nuclear blast in the Middle East.

Sodom and Gomorrah

Proof of Vitrification atop a mound in the Middle East

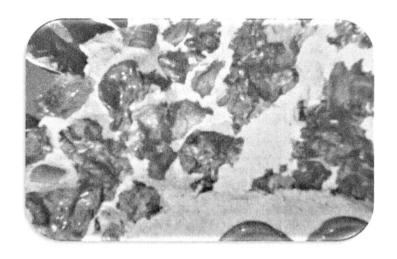

Glass from Vitrification

NUCLEAR BLASTS PRE-MODERN ERA

The clans of Enlil and Enki were apparently adversarial. The story of the Tower Of Babel may have been a war between the two clans. There is also evidence of a nuclear blast in India (before the "first" nuclear blast in Japan), which may have been the result of a "cosmic war." When asked during a question-and-answer session at a conference on atomic weapons research in 1952 whether the atomic weapon that had just exploded in Alamogordo, New Mexico, was *the first* ever to be detonated, a startled Robert Oppenheimer (father of the A-bomb) replied, "Yes it was the first one … **in modern times of course**." It is known that Oppenheimer was very well versed in Indian texts and *knew* that this *was not the first time* a nuclear blast had occurred on earth! He, in effect, admitted that this blast was the first "in modern times" (meaning it happened in prior times), of course. We didn't know this, because it was never taught to us in school. We were led to believe that the first explosion of a nuclear weapon happened *only* at a test in New Mexico and for use in war in Japan.

Sumer, which appeared six thousand years ago, vanished from the aftermath of this nuclear blast, becoming Babylon and Assyria. The Sumerian scholar Samuel Noah Kramer translated the Sumerian text as follows:

On the land [Sumer] fell a calamity, one unknown to man: one that had never been seen before, one which could not be withstood. A great storm from heaven … A land annihilating storm … An evil wind, like a rushing torrent … A battling storm joined by scorching heat … By day it deprived the land of the bright sun, in the evening the stars did not shine … The people, terrified, could hardly breathe; the evil wind clutched them, does not grant them another day … Mouths were drenched with blood … The face made pale by the Evil Wind. It causes cities to be desolate; stalls to become desolate, the sheepfolds to be emptied … Thus all its gods evacuated Uruk (Marrs, *Rule*).

It was at this time that the detailed narratives of Sumer and its gods ceased.

Edgar Cayce, the renowned psychic spiritualist said, "Just as the misuse of their spiritual powers brought turmoil, strife and questioning among themselves, men's misuse of scientific and material achievements [nuclear?] brought physical destruction in the earth. The destruction of the first world by volcanic action and fire. Some people were saved by hiding in caves and eventually emerged to begin a second world."

Is it also possible that our current symbol of medicine—entwined serpents—is a symbol of Enki. He engaged in genetic engineering to bring about Adam … and the entwined serpents represent the DNA double helix. So if a serpent from the Garden of Eden represents Enki, and Enki was the genetic scientist that "made" Adam, wouldn't it be reasonable to assume that the symbol of

medicine—two entwined serpents in the shape of a
DNA helix—could (and *should*)
represent our maker and designer of life?

This is really the key component to my book: that man is
not made from actual dust from the ground, and Eve is
not made from Adam's actual rib. It is that ETs from
another planet came here to create a slave race of
humans to help them get the raw materials they needed
from the earth, and, in the process, created humans to
achieve their goal. They used their own DNA and
combined it with a Cro-Magnon primate (*Australopithecus*)
to create a smarter bipedal primate—us.

CHAPTER 8
HUMAN EVOLUTION

Man has been conditioned for millennia to deny the truth of his ancestry and as a palliative we have developed a convenient form of amnesia. We have accepted the interpretation of history propagated by a self-perpetuating priesthood and academia.

--R. A. Boulay

"Ultimately the Darwinian theory of evolution is no more nor less than a great cosmogenic myth of the 20th century."

--Denton, Evolution: A Theory in Crisis

THE ORIGIN OF CHARLES DARWIN

1785—James Hutton publishes *Theory of the Origin of the Earth*, which stated that the earth was very old, and fossils represented the model with which to date different species. This is very important. This book uses fossils as *the* origin marker. It's *the* basis for physically proving our past.

1809—Jean-Baptiste Lamarck published *Philosophie Zoologique,* which classified different species according to a "ladder of nature." Ascending form the simplest to the most complex. He also suggested that the characteristics obtained during an organism's lifetime would be passed down to its offspring. Thus there is a "pyramid" of species, and the species passes down its learned traits genetically to its offspring.

1859—Darwin, who coincidentally was not a professional biologist and did not have either a doctorate or advanced degree in biology, publishes *The Origin of Species*, which extrapolates that all life evolved from a single source through the system of *natural selection*, and then, afterward, set out to prove his theory. His theory contradicted the theory of creation, supported by the church, which of course postulates that God created first Adam and Eve, who then multiplied. The core idea was that life evolved from a lower life-form, a "primordial soup." He endeavored to study animals and seek evolutionary evidence to support his theory. However, during the course of his study, he could not prove the "missing link" between man and primates.

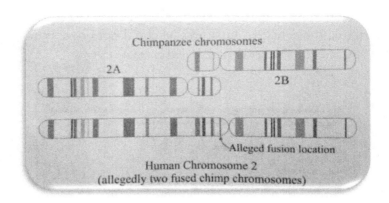

Fusion Marker showing a genetic manipulation in our chromosomes by an external force that intentionally modified the chromosome of a primate.

WHY DARWIN WAS WRONG

Taken from *Descent of Man,* this is Darwin's own *admission* of his mistake:

"I now admit … I probably attributed too much to the action of natural selection or the survival of the fittest. I had not formerly sufficiently considered the existence of many structures which appear to be, as far as we can judge, neither beneficial nor injurious; and this I believe this to be one of the greatest oversights as yet detected in my work."

He *admits* holes in his theory. This part we never read in our school books. Darwin has been glorified by the scientific community's as having solved the age old question of our existence. But his theory *cannot be proven* by the fossil record. There is no direct link between primate and humans. The direct link is "missing." So can his theories be true?

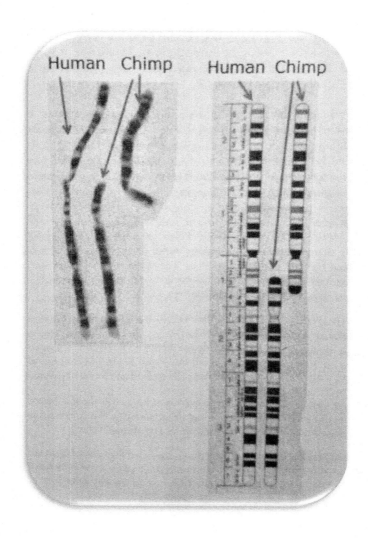

Our genome's code was broken in 1950 and co-discovered by Dr. Francis Crick. It contains 30,000 genes. Human beings are separated from chimpanzees by only 300 genes (1 percent), 223 of which do not have *any* predecessors on the genomic evolutionary tree! So where do these 223 genes come from ? Dr. Crick, and his co-discoverer, astronomer Fred Hoyle , have published criticism of Darwin's theory and have presented alternate theories to consider. Crick challenges Darwin on one hundred points, and suggests that "panspermia" not mutation and survival of the fittest , is the origin of species. Crick proposed that life was seeded on earth from spores that arrived (by a comet or planet) from outer space.

The problem with Darwin's theory is this: "If life slowly evolved from simple to complex forms through a series of mutations in response to changing environmental conditions, as Darwinism claims, then there should be superabundance of these mutated, intermediate forms. Yet the fossil record does not bear this out" (Hart). Thus, there are no transitional fossils. None. If Darwin were right, where are the fossils? Where is the proof? The funny thing is scientists *know* this, so they developed a theory called "punctuated equilibrium" which was developed by paleontologists Eldredge and Gould , which postulates that evolution occurs in short bursts of rapid change and is followed by long periods of stability. They then say that these spurts produced small populations in isolated regions of the planets and that these species died off without leaving a fossil record. Nice try.

So there are no fossils because the changed species were only thriving in "isolated regions" on earth? What about the non-isolated regions? Sounds interesting, but it provides no proof for Darwinism. It is a cover story to protect Darwinism in our classrooms books and continue the fallacy that *Homo sapiens* were naturally evolved and mutated primates.

Even Gould *admitted*, "All paleontologists know that the fossil record *contains precious little in the way of intermediate forms*: transitions between major groups are characteristically abrupt" (Hart).

Fred Hoyle also disagreed with the concept that life evolved from random mutations: "In short there is not a shred of objective evidence to support the hypothesis that life began in an organic soup here on Earth … Life did not appear by chance."

The transition from *Australopithecus* to *Homo Sapiens* was dramatic and not a slow evolution. Their brain capacity went from 500 cc to 750 cc literally overnight! This fact trumps the Darwinian concept of improvement/mutation, as it was way too quick and radical. There was an *intervention*, and I believe that the intervention was spawned by the Anunnaki , specifically Enki.

This was a jump, not a slow evolution:

Homo habilis—2.2 million—1.6 million years ago—average brain size—750 ml

Homo erectus—2.0 million—400,000 years ago—average brain size—900 ml

So it took 1.2 million years for our brains to enlarge from 750 ml to 900 ml (*an increase of 16 percent*)

Homo Sapiens—200,000 years ago—average brain size—1400 ml

How then, in 200,000 years, did our brains *jump to 55 percent larger*? Darwin said that we evolved very slowly and not in jumps. So how did we then jump to such a larger brain so quickly?

If panspermia is true, and our evolution did in fact jump quickly, then it all makes sense. Viruses and bacteria that arrive on earth from space infect our cells and cause mutation, thus causing dramatic DNA changes.

Neanderthals, however, are not related to humans, even though their brains were bigger--1500 cc compared to our 1400 cc capacity. They have been living on earth for 170,000 years, from 200,000 years ago to 30,000 years ago. They buried their dead and worked with tools but are not direct relatives. This has been proven by DNA analysis of a minute sample of a Neanderthal specimen. According to Lloyd Pye (one of my favorite authors), "378 base pairs of that ancient individual were compared with humans, and 28 mismatches were found. Among humans, that stretch of base pairs normally contains only 7 or 8 such mismatches, while chimpanzees have around 55. This places Neanderthals near the genetic midpoint between chimps and humans." He goes on to say, "For now we can leave the Darwinists scrambling to plausibly explain how 4 million years of supposed 'evolutionary' progress could so clearly point to the development of a robustly built, massively muscled, rather dimwitted brute; then suddenly, at only 120,000 years ago, modern humans appear as if by magic!--to live alongside their 'cousin' while looking almost nothing like them?"

Fossil discovery in Israel indicates that Neanderthal man lived side by side with Cro-Magnon man, yet the two races did not interbreed. According to Marrs in *Rule By Secrecy*, "Neanderthals and Moderns did not interbreed in the Levant because they *could* not. They were reproductively incompatible. Furthermore scientific testing of modern human remains show that they pre-dated the Neanderthals buy as much as 40,000 years! This presents a severe blow to the theory of continuous evolution. This proves that there is no "missing link" between primates and modern humans. They were two separate species. Something interfered.

Cro-Magnon's, on the other hand, *were* our predecessors. They appeared around 120,000 years ago and expanded in population at 35,000 years ago, while the Neanderthals became extinct at 30,000 years ago. Their skull shape was very similar to ours; they had high foreheads, and their brain capacity was 1400 cc. Their cave art from 20,000 years ago show humanlike artistic expression. Their final transition to becoming fully human occurred around 12,000 years ago.

Humans and gorillas share 98% genetic duplication. A human has forty-six chromosomes, while a gorilla has forty-eight. There are roughly 100,000 genes in a human's forty-six chromosomes, and about three million bits of DNA within those 100,000 genes, so the 2% difference produces sixty million base pair mismatches. To successfully cross-fertilize the two species, the genes *had to have been altered*! In order to do this, the Anunnaki used female eggs and combined two chromosomes into one by genetic manipulation (splicing and fusion).

"Man is not a descendant of the primates, nor is he under the protection of an Almighty God. We have tried both of these belief systems, and both have failed us. The systematic debunking of these theories is creating a crisis in both science and religion. It is time for scientists to repudiate Darwin and openly address the irrefutable evidence that man is an alien" (Conway).

"Then God said, 'Let *us* make man in *our* image, in *our* likeness" (Genesis 1:26).

Herein lies the smoking gun! The Bible we know has been written and then rewritten hundreds of times. Every time it was rewritten, it was to accommodate the desires of the ruling power at hand, whether through force or religion. This particular ruling clan--I believe it was the Hebrews in the Old Testament--had decided that it was *one* God- and changed the essence of the message, the essence that it was *more than one person* (gods) that made man in their image. "Our" refers to a group, not a singular person. This feeds exactly into the Sumerian story of the Anunnaki and their genetic manipulation of primates into *Homo sapiens*. If it was just one God that made man, it would have read, "I will create man in my image and my likeness." But it doesn't ... and even the scribes that change the story knew this, and didn't change the quote.

HUMAN EVOLUTION

The current view of human evolution pits divine creation against Darwin's notion of natural evolution by a process of random mutation and survival /selection of the fittest. All major religions reject the possibility that mankind and apes have a common ancestor. In the process of proving Darwin's position, scientists began to piece together fossil fragments in an effort to show that there had been a gradual development of a new species from an older one. They began to "date" fossils to fit Darwin's assumption. The late 20th century's use of DNA testing would later undermine these "dates."

Based on Lloyd Pye's fascinating book *Everything You Know Is Wrong*, his convincing argument is that human evolution does not stem from apes directly, through survival of the fittest (Darwin), and ordinary genetic mutation. *It was compromised. It was intervened with. It was manipulated. It was changed.*

Take for example the "missing link." The missing link is proof positive that man, while having 99% of a primate's genetic code, does not have 100%. Primates have forty-four chromosomes. Humans have forty-six. Now here's the kicker--According to Pye, *two chromosomes are fused together!* I repeat , *fused together*! This, if true, proves that man is an "invention" and a manipulated species derived by primates for the sole purpose of being slaves to ETs who needed gold for their dying planet. At the Max Planck Institute for Evolutionary Anthropology in

Germany , scientists have identified the gene FOXP 2
that gives humans control over muscles of the mouth
and throat. They reported that this gene mutation may
have occurred 200,000 years ago (Coincidence? See my
evolutionary chart), and that *this gene* differs from that of
chimpanzees by just two molecules!

According to Paul Von Ward in *We've Never Been Alone,*

Evidence of early humanoid fossils four plus million years ago suggests that the genus *Homo* COEXISTED with pre-humans and DID NOT genetically descend from them. This lack of intermediate fossils between those considered to be pre-human and early human can be described as the first missing link. It appears that members of this general erectus group overlapped with *Homo sapiens* for what could have been several hundred thousand years. Such overlaps suggest that one did not directly evolve into the other by a process of natural selection. We have *not* found fossils to fill the gap between the erectus types and *Homo sapiens.* A 1997 DNA study of Neanderthal remains (using a small amount of MtDNA)--concluded that modern human ARE NOT *direct descendants from Neanderthals.* DNA research has an edge over the fossil record. A truism among geneticists goes, "Genes always have ancestors while fossils may not have descendants."

FALSE TEACHINGS

Since I was a child, I had been taught the following in school:

1. The big bang created the universe thirteen billion years ago.
2. The Universe is expanding.
3. God created our Universe in six days and rested on the seventh.
4. Man is the most intelligent species.
5. There is no evidence of life on other planets.
6. Slaves built the pyramids for burial or religious purposes five thousand years ago.
7. Nothing travels faster than the speed of light.
8. Human evolved from primates.

Most people would tend to agree with many of the above statements. They have been taught in their schools or by their parents or guardians. They have been formed in people's minds as *fact*, although they are unsupportable by available evidence! In his book "Uncommon Knowledge," Al McDowell cites several points that contradicts our "schoolhouse truths."

1. In his book *The Big Bang Never Happened*, Eric Lerner points out that based on the current clustering of our galaxies, the universe would have to be one hundred billion years old, not thirteen billion years old. He argues that the universe has been recycling through dissipation of energy and mass from stars. Thirty-three scientists published objections to The Big Bang

theory in 2004, and a conference was held in Portugal to discuss cosmological evidence in opposition to the theory.

2. There is the belief in an expanding universe in which all galaxies travel away from our galaxy at a proportional velocity to their distance from us. Edwin Hubble discovered, in 1929, that fainter galaxies had slightly longer wavelengths of light arriving to earth. This increase in light wavelength is called redshift. Halton Arp (who worked under Hubble) has shown that light in fact *decreases* and loses energy as it travels.

3. Mitochondrial DNA studies indicate that mankind is 200,000 years old. I believe that the Anunnaki ETs from Nibiru colonized earth 400,000 years ago and brought with them multicellular organisms from Nibiru to earth from their first arrival to all subsequent arrival and/or cataclysms. Archaeological remains of early humans conclude that human-like life existed prior to the Anunnaki, but they were not *Homo sapiens*.

4. The Dogon tribe in West Africa make claim that the earth originated from the Sirius solar system (a nearby visible system in our sky). The Dogon's say that they originated in Egypt from Sirius , the Dog Star, which is only 8.6 light years away from us. The Dogon say that the Sirius system has three suns: Sirius A (visible to us); Sirius B, that is a nearly invisible white dwarf; and Sirius C, which is a

habitable planet. Dogon's say that Sirius B rotates on its axis like earth and has a elliptical orbit of fifty years. Until only recently, scientists were unaware that Sirius was a multiple-star system, and with all of the Dogon's information that they "learned" coincides with what scientists have recently discovered. How could they have known that information? Coincidence?

5. While I love Einstein, and believe he is responsible for our incredible growth of knowledge in the sciences, the underlying premise of Einstein's special theory of relativity may not be accurate. Light *does not* have a constant speed in a vacuum as seen by observers in different frames of reference moving with respect to each other. Instead, evidence suggests that light is conducted through space in a medium never discussed in our classrooms called "the ether" and that the speed of light is *determined* by this ether. This ether theory of light contradicts the current scientific presumption that nothing travels faster than the speed of light. This invisible ether has the ingredients of gravity, electricity, and electromagnetism. It exists within our own dimension of space and time and is completely governed by Newtonian mechanics.

6. There are two genes that separate humans from primates—MYH16 and FOXP2. MYH16, studied by Dr. Hansell Stedman, regulates cranial development. Cranial structure was modified from an oversize jaw and small brain found in primates, to a smaller jaw,

larger brained early hominids. Humans differed from primates by 950 cc of brain size. It appears as if the older form of MYH16 was "snipped out" and replaced with a new, modified form which was then "tweezered back" into the genome of an early ancestor. So this *was not* a random mutation--this was an intervention. Darwin suggested that life evolved in a "warm little pond" with haphazard ingredients thrown together, mixing like a sludgy soup. In 1857, Louis Pasteur proved that microorganisms derived from pre-existing microbes, so that *our life* derived from life that existed before. So, in essence, Darwin was wrong. Life did not begin haphazardly; it evolved from prior life. Unbelievably, what I have been taught and believed in for years is a disproven theory! Our schools in the United States *still* teach Darwin's theories and promote them as fact.

When DNA was discovered in 1953 by Dr. James Watson and Francis Crick , the world could see how incredibly complex the DNA molecule was. The human mind can store terabytes of information on what looks essentially like grand of sand! There was *no way* that a DNA strand could have evolved in a "warm little pond." "At a molecular level, life even in its primitive form, is so incredibly complex that any prospect of transforming an inorganic system into biology must be considered awesomely difficult to say the least." (Coppens). The point is that there was not enough time on planet Earth for such a complex system like a DNA molecule to form here. Crick determined that DNA had come from elsewhere in the universe, thus creating the theory of panspermia--that our DNA is an accumulation of material from comets passing through our solar system, one or more of which landed on earth. Was our universe somehow designed to create RNA/DNA-based life?

WHAT ARE WE?

Here are some interesting quotes from well-known authors in the field:

"What actually transpired," explains Richard Gardner, "was that the original Mesopotamian writings were recorded as history. This history was later rewritten to form a base for foreign religious cults--first Judaism and then Christianity. The corrupted dogma--the new approved history--was so different from the original writings, the only first-hand reports were labeled 'mythology.'"

William Bramley quotes, "Human beings appear to be a slave race languishing on an isolated planet in a small galaxy. As such, the human race was once a source of labor for an extraterrestrial civilization and still remains a possession today. To keep control over its possession and to maintain Earth as something of a prison, that other civilization has bred never-ending conflict between human beings, has promoted human spiritual decay, and has erected on Earth conditions of unremitting physical hardship. The situation has lasted for thousands of years and it continues today."

R. A. Boulay opines, "Man has been condition for millennia to deny the truth of his ancestry and as a palliative we have developed a convenient form of amnesia. We have accepted the interpretation of history propagated by self-perpetuating priesthood and academia."

Charles Fort in 1941 concludes, "We are property. I should say we belong to something; that once upon a time, this Earth was No-Man's Land, that other worlds explored and colonized here, and fought among themselves for possession, but now its owned by something."

CHAPTER 9
ADAM AND EVE

ADAM

As a young boy, I remember learning about Adam--a tall, muscular, Caucasian young man. The name Adam made sense. It used the first letter of the alphabet, it's not too long, not too short. He walked around the garden of Eden barefoot with a fig leaf covering his private parts. What if Adam was not an individual at all? What if Adam was but a race of ET men? What if Eve was a race of intelligent, upright-walking primates?

The Anunnaki called the first humans "Adamu"—no coincidence here. Adamu minus the *u* equals Adam. So what if Adam wasn't really Adam at all? What if Adam were many men? These men became hybrids through the genetic manipulation of Anunnaki and primate DNA, and then these men had intercourse with a group of females going through the same manipulation so that Adam and Eve was the first trial group of protohumans, the first group that passed muster that were not primates and were not Anunnaki.

I struggled with the understanding of Adam and Eve eating from the Tree of Knowledge and its relation to good versus evil. What good, and what evil? Was the good "knowledge" separating man from beast? Was the good "sex" with Eve? Was the evil "knowledge" that humans were naked? Was the evil "sex" with Eve? Was Enki the "snake" that lured them?

What knowledge did Adam gain? Was it the knowledge of the Anunnaki? The Bible states that Adam and Eve suddenly had the awareness that they were naked. Was this the knowledge? They then clothed themselves with fig leaves. This makes sense because figs were readily available in the Tigris-Euphrates region, where as apples were not! Apples were not present whatsoever in the region. It's apparent to me that the "apple" was substituted for "fruit" in one of the many rewrites of the Old Testament. And trust me when I tell you, there were *many* rewrites!

Why then did Adam and Eve get evicted from the garden of Eden? We know from the Bible it was a result of being coaxed by the snake to eat the fruit from the Tree of Knowledge. The snake promised them equality with the "gods" if they ate, right? So eating the fruit would give them what equality exactly? Knowledge or immortality? I think it was both. I think that Adam and Eve saw that the Anunnaki were vastly superior and seemingly immortal. Human nature kicked in, and they wanted to be like their parents. They wanted to be gods.

So, basically, just as the title of my book reads, Adam *was an alien*! He was also not a singular man. He was many men. It was the final result of the Anunnaki experiment to produce a human slave. He was whom we read about as children as *the first* human male. Eve was also an alien. She was the result of the Anunnaki experiment as well.

I also think Adam was not Caucasian, but dark skinned. There are many references to "the black heads" in the tablets. The "black heads" were a reference to the human slaves in the South African mines. Why "black heads"? I think because their shared genetic makeup with primates, that they were mostly dark skin (maybe similar to present day aboriginals). And they were considered "black" to the fair skinned, blue-eyed Anunnaki.

So why then were they evicted? It's my belief that their "jealous gods" wanted servants, not more gods. It just feeds into the notion that humans were born as servants purely to satisfy the needs of their gods--whether it be gold, silver, or helping to build extravagant monuments to them.

The whole Cain-and-Abel saga reinforces the good/evil theme , with Cain killing Abel because of jealousy. Cain is then spared by God, and his "punishment" is, once again, being pushed out of the Tigris- Euphrates valley. (Isn't that interesting? The gods punish humans by banishing them from a certain area. We find out later that their punishment does in fact include death.) He then "builds a city" and "fathers children" (Genesis 4:17). So there's the million-dollar question: with whom did he (Cain) father children if not with Eve? If Adam and Eve where the first two humans who fathered Cain, Abel, and later Seth (who was born after Cain kills Abel), with whom did Cain father children? Cain was then banished to the "land of Nod." So there now are two men , one woman, and one male child. Cain's child name is Enoch. With whom did he have this son, Enoch? Seems to me obvious that Cain fathered Enoch with an Anunnaki woman. Who else could it have been?

Wonderful Sumerian depiction of the birth of Adam. The jars represent the receptacle of sperm and egg, or a "test tube baby". Adam sits on the lap of his Anunnaki maker.

The tree of life behind them.

EVE

When Eve eats the fruit from the forbidden tree, she was beguiled by the snake to do so. Previous reading tells us that the snake is Enki. Let's think about the words that the author of the rewritten Old Testament uses:

"tree"—not a physical tree, but it may refer to a "family tree"

"fruit of"—not an apple, but it may refer to Eve's genitals

"snake"—not an actual snake, but it may refer to Adam's genitals, or Enki

"eat of it"—not eating an apple, but it may refer to the act of intercourse

The Old Testament claims that the serpent, had arms and legs and walked upright. After having intercourse with Eve, the ministering angels descended and cut off its arms and legs—"Upon thy belly shall thou go" (Beresheit 3:14).

My take on this is that Eve had intercourse with an Anunnaki *before* she had intercourse with Adam. In essence, she had committed adultery. She lost her virginity to the serpent (Enki) and became impregnated by it. Thereafter, Adam had intercourse with Eve. Maybe she subsequently became pregnant with Cain and Abel. Maybe symbolically the twins represented the origin of "good and evil." "Good" being Abel, the son of Adam, and "Evil" being Cain, the son of Enki.

Was the author conveniently using twins to represent conflict? To represent the good-versus-evil dogma? To initiate conflict among humans right away ? These opposing lineages would be in constant conflict with one another, resulting of course in Cain murdering Abel. Cain was the first of the *Nephilim,* which is used several times in the Bible.

In the book of Enoch, the Watchers are angels who have been dispatched or assigned to the earth to "watch over" its inhabitants, but in the process of conducting their duties, they became enamored with human women and conceived offspring called Nephilim. It's also interesting to note about the "angels" who have been depicted with wings since the Middle Ages. Why wings? The contemporary reasoning is that they come from "heaven", and heaven is above earth. So they "flew" down, right? Wrong! Nowhere in the Old Testament or New Testament does it say that angels had wings. They "fly" but do not have wings. It was a Christian artist who painted them with wings in order to support the flight theory. So if they flew without wings, how did they get to the earth from heaven?

And that's where UFOs come in ...

CHAPTER 10
NOAH

The Book of Noah discusses the origin of his birth. He was the grandson of Methusaleh, and the son of Lamech. Lamech stated that he had suspicion that his wife had been impregnated by one of the Watchers. Enoch was Methuselah's father and was already integrated with the ETs. According to Von Ward, "The exchange between the two gives *prima facie* evidence that Noah was a hybrid. Methuselah told Enoch that Lamech had said his son was 'unlike man, and resembling the sons of the God of Heaven.' Enoch responded by saying that in the time of his father Jared 'some of the angels of heaven ... united themselves with women and have begotten children by them.' Lamech describes his wife's son as having skin 'white as snow.' Yet 'red as the blooming of a rose.' His hair was long, wooly and white. His eyes were beautiful and bright [perhaps blue?]. Enoch predicted that Noah would be one of the 'giants on earth.'"

Christian and Barbara Joy O'Brien, a British husband-and-wife team of researchers, studied the technological revelations of the Torah (Jewish Bible) and its relation to YHWH (the Hebrew symbol for God). They believe that "the Shining Ones" as depicted in the Torah were an "advanced, genetically related race to mankind, who [also] possessed an advanced technology which, in

Yahweh's [YHWH] case, they believe was used to cow and coerce the Israelites into unhesitating, unquestioning obedience to Yahweh and his plans for conquest" (Farrell, *Genes, Giants, Monsters, and Men*). And, "before we discuss the enigma of Yahweh proper, it must be stressed that the building of the Hebrew nation was achieved by a deliberate process of selection which stretched back to Noah. His family was selected for survival after the Flood; and Terah's family was selected from the Semitic people of Ur; and out of it Abraham was chosen, over his brothers, for reasons that we are unsure. Possibly they were genetic; but another aspect may have been that Sarah, Abrahams wife, was barren-- and this may have given the opportunity for another remarkable genetic opportunity" (Farrell, Genes).

The O'Brien's study of Yahweh separate Yahweh from a simple human by commenting that Yahweh always kept his face hidden from humans. Why? Was he gruesome? Ugly? It appears that when Yahweh communicated with Abraham, his face was always covered. Did Yahweh carry some sort of bacteria that could infect humans? However, he was a physical being, not a spirit. He was probably of exceptional size, taller than typical humans at the time and considered a giant.

There are also correlations of the Shining Ones, taken from the three principal sources:

- Sumerian tablets in which they were referred to as The Anunnaki;
- The Hebrew book of Enoch in which they were described as Angels; and

- The book of Genesis in which they were referred to as Elohim

According to the Torah, "The Nephilim were on the *earth in those days—and also afterword*—when the sons of God went to the daughters of humans and had children by them. They were the heroes of old, men of renown." Consider the words "in those days" and "and also afterword." Clearly, the scribe is saying that the Nephilim (ETs) came and went. When they came, they procreated with the female *Homo Sapiens* on earth and begat children.

It's also interesting that when the Torah and the Bible discuss wanting to "shed blood," it is obvious to me that it is for purposes of genetics and not to kill. Maybe the Anunnaki wanted to check to see who was the most able and suitable for a particular task.

Regarding today's religions, the O'Briens break down two possibilities of the genetic constitution of mankind:

- If Cro-Magnon man were a hybrid of Neanderthal man and Anunnaki, and the patriarchal tribes were hybrids of Cro-Magnon man and the Anunnaki, then the patriarchal tribes people—who were the progenitors of the Jewish race—were three-fourths Anunnaki and one-fourth Neanderthal.
- Alternatively, if Cro-Magnon were not a hybrid, but an evolutionary mutation of early man, then the patriarchal tribes were one-half Anunnaki and one-half early man.

- In either case, it has been stated that members of the Jewish race, through their patriarchal progenitors, carry more of the "divine" Anunnaki strain in their cells. This is only a possibility of why the Jewish people are considered "The Chosen Ones" but is only a postulation.

They cite Noah as a perfect example of a hybrid: "His body was as white as snow and red as the rose, and the hair of his head was in long locks which were as white as wool; and his eyes were beautiful" (Farrell, *Genes*).

THE FLOOD

Noah was reportedly warned of the impending flood by Enki and was given exact specifications to build the ark by Enki as well. After the flood, the ark is purported to have beached on Mount Ararat in present-day Turkey. Noah was considered by the Anunnaki as a "special" human, probably more than 51 % genetically an ET and very wise. His three sons, Shem, Ham, and Japheth, also carried these genes.

According to the Bible, Noah's ark was 450 feet long (300 cubits), 75 feet wide (50 cubits), and 45 feet tall (30 cubits). A cubit is 18 inches. We are told that it was made of wood. Is this possible? The longest wooden ship ever built in the modern day is the *Wyoming* which was 329 feet long. It required iron strapping for support and leaked badly. Noah's ark also had no rudder. So is this ark possible?

"And of every living thing of all flesh, two of every sort shall bring into the ark" (Genesis 6:18--21). So Genesis tells us that Noah's ark *carried* one of each species of all terrain animals on earth? In 450 feet? Impossible! Is that even remotely logical? Let's say that 1000 species of land animals existed at the time, and we must include elephants , giraffes, rhinos, and so on ... They were *all* on that Ark? As my daughter would say, "No way!"

Even *if* the ark were 450 feet long, what about animals that were not in the Tigris-Euphrates Valley? Also, the rain for forty days and forty should not to be taken literally. In the Bible, the number forty represented "a whole lot". Words used in the Bible need not be taken literally. They were the author's use of words to cover the truth or to disguise real events that proved impossible.

What if the reality was that no actual, physical animals were on the ark? What if Noah carried the **DNA of animals only, as in eggs and sperm**? Not sure about this idea … Just know that a 450 foot long wooden boat cannot physically carry all living land animals that existed in the Tigris-Euphrates valley.

But why did the flood occur? Was it initiated by a meteorite? Was it initiated by a massive global warming that melted the polar ice caps? Was it done intentionally by the gods to wipe out their human species, maybe because humans were growing violent to one another, and start anew? Here are quotes from Genesis: "Then the Lord saw that the wickedness of man was great on Earth, and that every intent of the thoughts of his heart was only evil continually," "I will blot out man whom I have created from the face of the land, from man to animals to creeping things and to birds of the sky, for I am sorry that I have made them" (Genesis 6:7).

So the quote above is an admission from the Anunnaki that their creation was:

- not created correctly;
- disobedient; and
- too intelligent.

Stories like the Tower of Babel and the destruction of the Great Temple in AD 70 make me believe that the ETs tried to wipe out their mistakes not once but three separate times: first, at the initial test phase of making humans; second at the Tower of Babel; and third at Solomon's Temple. Three separate attempts, the second and possibly third of which included the ETs use of nuclear weapons as evidenced by vitrification of rock on earth.

CATACLYSM

There is considerable evidence to suggest that a cataclysm occurred 11,500 years ago. We have all read about Noah's Ark. This is a direct result of such a cataclysm. Some suggest that prior to the cataclysm, the earth's axis was more perpendicular to its orbit (with fewer seasons), the planet was hotter with more flora and fauna, and that it had smaller polar caps, lower mountains, and shallow seas. The atmosphere had more carbon dioxide and oxygen. The geological record shows that the earth has been subjected to several cataclysms over the last five billion years, several of which occurred during the lifespan of *Homo sapiens*.

Noah's tale described in chapter 6--8 of Genesis had a cataclysm which lasted for "40 days and nights" with constant rain. The waters swelled, covering all of the mountains and lifting the ark "to rise 15 cubits above the mountains." The Hebrews saw the flood as punishment by the Lord (Enlil) for man's evil thoughts.

My research has concluded that there have been four cataclysms:

- 7,900 years ago
- 4,300 years ago
- 700 years ago

THE GREAT FLOOD

- Interesting also is the sudden explosion of plant and animal life right after the flood 11,500 years ago. Typically, with any planetary disaster, life would be decimated, and surely species of animals were. However, there appears to be evidence that there was an explosion of life after the flood, allowing for the human diet of plants and agriculture. Corn, for example, has no wild ancestor and *could not have evolved from earth.* Natives in Mexico praise corn as a "gift from the gods." Does this explosion suggest an ET intervention?

- 11,500 years ago--The flood. Was this caused by a passing comet, asteroid, or planet? Is it possible that a near miss could have created the devastating effects of this flood? In ancient Greek mythology, a planet called Phaeton was pulled into our solar system's gravity field and caused the first cataclysm. The Sumerians attribute it to the Nibiru. Babylonians called it Marduk. Is it possible that three separate civilizations could or would label a planet-like body as the primary cause of the flood? Is it also possible that our Kuiper Belt (asteroid belt) is the *direct result* of this planet like body smashing into a former planet in our solar system? What then explains the asteroid belt fragments?

- The impact could have created a shift in the earth's rotational axis (from 30 degrees to 23.5 degrees) and caused volcanic and seismic activity, hurricanes, and flooding. The sudden extinction of the wooly mammoths and saber-toothed tigers could have been attributed to this. Evidence of frozen wooly mammoths found with undigested food in their stomachs contributes to this theory. Even Edgar Cayce (the famous psychic) reported in his reading #1681 that "the Atlanteans had been warned by the Advanced Beings [about the flood] who flew in vehicles over the earth" (Von Ward). Machu Pichu may be a direct result of the humans building cities atop high mountains to escape from a recurrence of floods.

MS 3026
The Sumerian Flood story, Babylonia, 19th-18th c. BC

Sumerian Tablet—The Story of the Flood. It describes
when, where and why it happened.

THE ARK OF THE COVENANT

- The Ark of the Covenant is perceived by most religious people as the ornate container in which the Jews carried the Ten Commandments.

- But there are many references to the ark being deadly, having killed humans for opening it or "using it" incorrectly. How would it actually kill someone? There had to be something special about it, not just that it was sacred. What was it?

- I believe That the Ark of the Covenant wasn't just a chest that held the Ten Commandments but rather a communication device between the early Jews and Yahweh/Enlil, when he was away, to be issued commands. A modern reconstruction of the ark in the 1980s, built from the exact instructions found in the Bible, resulted in an electrical capacitor of powerful proportions. Author Paul Schroeder says enough evidence exists to suggest that the ark was created as an "alien transmitter-capacitor," and as a form of weapon. "One of the gold plates was positively charged and one was negatively charged and together they formed the condenser. If one of the cherubims positioned above the Mercy Seat acted as a magnet, then one has the rudimentary requirements of a two-way communication set" (Shroeder, *The Pyramids and the Pentagon*).

- J.C. Vintner in *Ancient Earth Mysteries* says that the ark might have been radioactive on some level and claims that those who touched the ark died instantly or shortly thereafter.

CHAPTER 11
PHYSICAL EVIDENCE
OF ETS

Antikythera Mechanism. The world's first computer, dated 100 BC, was found in 1900 off the island of Antikythera near Crete. The device contained a system of differential gears not used until the 16th century. Yale's Professor Price stated that it was a mechanical analog computer, an instrument millennia ahead of its time. It calculated the motions of the stars and planets. When past or future dates were entered via a crank, it calculated the position of the sun , moon, or other planets. Made of bronze and standing 13 inches high, it is inscribed with more than 2000 characters. It was originally found at the bottom of the sea in a sunken Roman ship and was used as a navigation device that permitted the ship to cross the Atlantic to America more than 1500 years *before Columbus*!

Baghdad Battery. The world's first battery, discovered by archaeologist Wilhelm Koenig in 1938, is dated back 250 BC! Discovered in an Iraqi village, this small vessel containing a copper cylinder with an iron rod produced electricity when alkaline grape juice was added to the vessel.

Stonehenge. Stonehenge, in Wilshire, England, was
built as far back as 8000 BC, and this marvel of
construction defies human ability. It consists of sixty
bluestones, weighing up to four tons and is believed to
have originated in the Preseli Mountains 240 miles away.
Stones are lifted to create a "staple" shape, or an open
doorway perhaps. What was its purpose? A spaceport?
Time Machine? Energy device? The stones were formed
in the shape of a circle. I believe it was an ancient
astronomical calendar. Alignments between the massive
stones coincide with the summer and winter solstice, the
equinox, and the phases of the moon. The number sixty
is the classic Anunnaki sexagesimal number which is also
embedded in our standard of time (60 seconds x 60
minutes = 3,600 = the number of years in Nibiru's [the
planet of Anunnaki] orbit). Coincidence?

Easter Island. Here in the middle of nowhere, eighty-
ton statues chiseled from volcanic ash and resembling
tribal human faces were built. The "Moai," as they are
called by the locals, are seven eighteen-ton statues
pointing directing to the sun setting during the equinox
and are named for the Seven Sisters of the Pleiades.

Nazca Lines. These are 6,900 man sized holes carved
into the ground, stretching the plains in Nazca, Peru.
From the ground they appear to have no purpose. But
from this sky, they pose something altogether different.
From the sky, they appear to be *runways* or *landing strips*!
How can this be? Built thousands of years ago? The
bands of holes are so precise that they resemble the work

of a drilling machine that produced holes in the earth for several miles.

Piri Reis Map. The maps of Turkish admiral Piri Reis dated in the early 16th century accurately depicted the northern coastline of Antarctica, which wasn't discovered until 1819. The only way to have depicted the Antarctic coastline in the 16th century to the accurateness of this map would have been by air . Antarctica was also covered by a glacier in the early 16th century. So who made this map? And how?

Baalbek Megaliths. In the Beka Valley in Lebanon, there exist structures that defy human possibility. Baalbek used to be called Heliopolis when Caesar conquered Syria in 47 BC. These solid limestone blocks are the largest and heaviest stones used in human construction known on earth! I repeat, the largest and heaviest stones cut on the earth! The largest stones are 1,200 tons or 2.4 million pounds! The threshold amount of weight that could be lifted by the Romans at this time was 200 tons. How can six times the threshold of human lift be possible?

Saqqara Bird. This object found in a tomb complex in Saqqara, Egypt, now in the Museum of Cairo, would probably go unnoticed. It appears to be a miniature hawk, but if looks closely, he or she sees the aeronautical design of a plane! Another figure found in Quimbaya in Colombia also resembles a bird, but a group of German aeronautic engineers led by Peter Belting saw something completely different. In 1996, working off measurements of the original bird, they constructed a larger-scale, sixteen-to-one model made out of balsa wood. It flew in a wind tunnel. This was no bird. It was a plane.

Nitinol. The "Roswell miracle metal" found in the crashed Roswell craft is not found on earth, and scientists are unable to duplicate it in labs. We know that it is a combination of extremely pure titanium and another metal unknown to man, combined in an irregular way.

Dropa Discs. Discovered in 1937--38, 716 metal, fifteen inch wide disk were found in the Kunlun-Kette mountain range by a Chinese expedition from Peking. Attempting to find shelter, the team members entered a cave only to find several tombs, aligned and rows, with skeletons with abnormally large skulls. Buried with the skeletons were these unusual discs with a hole in the center, each bearing strange hieroglyphs. The discs were deciphered in 1962 to be the account of a spaceship that descended to earth 12,000 years ago in the mountains of Baian-Kara-Ula. The ET's were chased and killed by members of the Han tribe, who lived in nearby caves. They were described as "short, skinny, yellow men with big, knobby heads and small bodies and were a terrible sight to see"(Coppens). Later Soviet scientist Dr. Viatcheslav Saizev placed the disks on a gramophone. When they turned on, the disk vibrated and hummed, essentially forming an electrical circuit.

Gobekli Tepe. Built in 9000 BC, these megalithic blocks are estimated to weigh seven to fifty tons each. Archaeologists discovered a small sphinx that looks similar to *the* Sphinx, which was replicated thousands of years later. This city in modern day Turkey was one of the first cities built *after* the flood.

Carnac. Over three thousand megaliths weighing between 20 and 350 tons each outside the French village of the Carnac in France. Carbon dating places the stones to be as old as 4500 BC, during the Neolithic Period. The stones are arranged in alignments that display mathematical concepts like the Pythagorean theorem, thousands of years prior to it being originated by Pythagoras!

Bosnian Pyramid. This structure is estimated to date to 32,000 BC! At 722 feet, it is the largest pyramid in the world. It dwarfs even the Great Pyramid at Giza. It has been named Visocica. Very little is published about this pyramid.

Nazca Lines in Peru. See the runways?

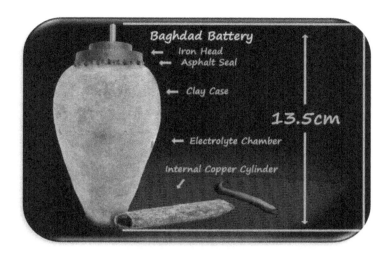

The Baghdad Battery

Coral Castle. Built by one man using nothing but
magnets and sound to levitate ton of rock.

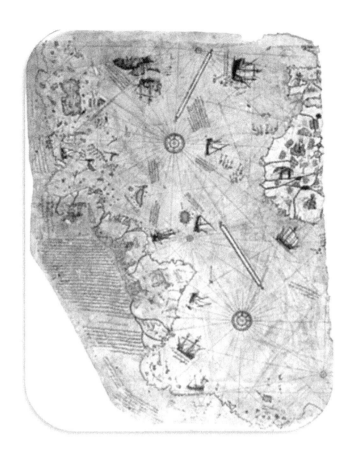

Piri Reis Map

Baalbek Megalith. This is the largest monolith in the world. It was made as a landing strip for Anunnaki Spaceships.

Saqqara Bird, thousands of years old.

Space Shuttle, present time. Coincidence of Design?

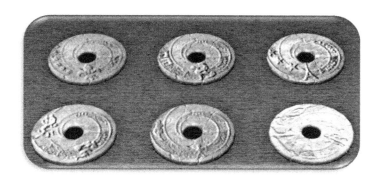

The Dropa Discs

CHAPTER 12
BLOOD

The Jewish people have been called "the Chosen Ones" to many 'religions' chagrin. For close to five thousand years, they have been the bull's-eye target of many religions' and world governments' wrath. Today, the Jews make up less than 0.25 percent of the world's population (only about 18,000,000 Jews live today, as of 2013),but they are responsible for 162 out of the 750 Nobel Prizes issued from 1901 to 2007:

- Chemistry—33 Nobel Prizes, or 20 percent of the total
- Economics—29 Nobel Prizes, or 41 percent of the total
- Literature—13 Nobel Prizes, or 12 percent of the total
- Physics—50 Nobel Prizes, or 26 percent of the total
- Medicine—53 Nobel Prizes, or 26 percent of the total

How can this be? Such a small percentage of the world's population, having contributed to 22 percent of all the Nobel Peace Prizes awarded? Notwithstanding areas of sports, philosophy, anthropology, and so on to which they have contributed. It is an inordinate percentage, and there truly needs to be an explanation for it other than "survival of the fittest."

But one thing that came up in my research that I found very strange was their blood makeup—Ashkenazi Jews, in particular. They have a unique Rh factor, or *delta-*CCR5 genes, as compared to other Jews. Why? Where did this gene come from? Why them?

The original human blood type is O. It is associated with meat eaters. Until 25,000 years ago did type A appear somewhere either in Asia or the Middle East. Type A blood is associated with eating grains and agricultural crops. Type AB (the author's blood type) is very rare and only came into existence 1,000 years ago . Type B apparently occurred just after the cataclysm of 11,500 years ago. Historically, Jews have higher-than-average Type B blood, while Arabs have higher-than-average Type O.

My point in discussing blood types is to demonstrate that the Anunnaki bloodline, may in fact, be attributable to man's remarkable accomplishments. Geniuses like Einstein , Tesla, Newton, etc. may have a higher concentration of Anunnaki blood then other humans.

Now, back to religion, after the Anunnaki exploded a nuclear device in the Middle East because its creation got out-of-hand, the Israelites were forced to move.

Abraham and his people moved away from the devastation to the South, where he fathered Isaac at the age of 100 years, thanks to his *hybrid* genes. Isaac's son Jacob came to be known as Israel, a name soon applied to his entire people. Some believe that the name Israel is nothing less than a combination of the Egyptian gods: *Osiris*, *Ra*, and *El*—Israel.

Growing up as a fairly religious person, I sought out the positives of it--family get togethers, closeness among siblings and cousins, learning right from wrong, respecting our parents and elders, and so on--but as my research progressed, my attitude changed. I felt that while religion does in fact provide very good values for people worldwide, binds them, and gives them purpose, it can seed evil as well. My biggest issue is, how do we *really* know that everything that is written in the Bible is *true*? How do we know? Where is the evidence? Where is heaven?

So my research into the **Sumerian tablets** took me off into another direction. **What if what it said were true?** It contradicts almost everything we learned through religion. But it does do one thing- it explains a real story that may have actually occurred and nudges us to look, really look, at the physical evidence purportedly built by humans under fantastic and impossible means. Today, so much of the world follows the letter of the law with religion that, besides killing for it, transcends their normal instincts. **It asks us to believe in the unbelievable**. It ask us not to question what we see. This is where I get *itchy* …

Any breakthrough in knowledge of our past, with evidence, is relegated to the esoteric or occult. It is laughed at, especially by religion. It is considered "evil" if it doesn't fit the story we are taught in elementary school, college, or Sunday school. How can Darwin be Wrong? How can UFOs really exist? Aliens are for crazy people! Why scare our kids about them?

My belief is that religion has held down the truth about our origins for millennia. Why not support a cause with an invisible God that receives donations from millions of people based upon sheer "faith"? And that faith believes that we are created directly from God, as humans, in the Garden of Eden. Perfect creatures of earth, with no competition in the universe. The one and only sentient beings. Rulers of the galaxy. The problem is not only that religion has masked the truth about our origins; its agenda is mental control and wealth accumulation. The Bible was written , passed down, and changed for millennia. Each generation of scribes that wrote the Bible threw something in that supported their agendas. Many of the rules, laws, and stories of the Torah (Jewish Bible) were actually recycled from the Babylonians and Sumerians during the Jewish exile from Babylon. The Ten Commandments were recycled from the Babylonian law called the Code of Hammurabi, which dates back to 1772 BC. The code of Hammurabi was recycled from the Sumerian Code of Ur-Nammi, which is the oldest known code, having begun in 2095 BC. It appears that Judaism developed largely in and around 200 BC, Christianity recycled from Judaism in AD 100-300, and then Islam from both Judaism and Christianity in AD 600-700.

There are many laws of religion that don't make sense, aren't supported by proof or fact, and depend upon one's blind obedience. When religion bleeds into government is when I have a problem. The Vatican publicly states that UFOs and extraterrestrial life exist but won't admit that the Sumerian tablets, written thousands of years before the New Testament, tell the real story of how human beings came in to existence. And I believe, they know it is true.

CHAPTER 13
ZECHARIA SITCHIN

*"You never change things by fighting the existing reality. To change
something, build a new model that makes the existing model
obsolete."*

--R. Buckminster Fuller

I owe much of my inspiration regarding the Sumerian
Tablets and the Anunnaki to a true warrior in the field –
Zecharia Sitchin. He was one of only 200 scholars that
could decipher the first known writing on earth --
cuneiform script. He could read Sumerian, Akkadian,
and Assyrian languages. He spent decades translating
them and comparing them to the Old Testament. Yet he
is unknown to the public. His work is far ahead of its
time and is under-appreciated, misunderstood, and
generally not accepted by the majority of mainstream
scientists. His thesis was that the stories written on these
tablets were *real life events* that truly occurred, and that
mythology was *based on* these true events.

*"What you thought was mythology is actually history. In many
cases, all those gods mentioned in myth after myth were real, and
they made a considerable impact on the emerging human race"* –
Jack Barranger

The theme that modern science is now, finally catching up with pre-existing, ancient events is very real. The existence of Nibiru (or Planet X) has been confirmed by U.S. Naval Observatory scientist Robert Harrington, in collaboration with astronomer and physicist Tom Van Flandern. They confirmed that it is, in fact, a very large real planet (3X-5X larger than earth) that orbits **our** solar system elliptically from deep space. Flandern is a specialist in celestial mechanics and is a very reliable expert to confirm this data. Planet X was published in the Washington Post on December 30th, 1983.

The following ancient text describe Nibiru as a large inhabited planet in our solar system that circles the sun in an extreme ellipsis—

- Enuma Elish
- The Atrahasis – the foundation for the first six chapters of Genesis
- Karsaq Epics

All texts describe Nibiru entering the circumference of our solar system every 3600 years. As it approached Earth, these advanced beings (The Anunnaki) would come to earth and colonize it. So here's my question-- why do most people believe that what's written in the Old Testament or The Bible is absolute truth, when these texts above PRECEED them! These texts were the BASIS of what's written in the Old Testament and The Bible. If we can believe that Moses parted the Red Sea or that Jesus walked on water, why not believe that advanced beings created us to begin with? What's so bad about it?

We should be thanking them!

CHAPTER 14
THE STORY OF OUR ANCESTORS THE ANUNNAKI

Here is what I believe happened, based on the Sumerian tablets.

1. "In the 1850s European amateur archaeologist began to dig in the mounds of the Tigris-Euphrates Valley (present day Iraq) seeking to uncover the ruins of the fabled cities of Mesopotamia" (Ward).

2. In the process of excavating the ruins of Ninevah and Nimrud, thousands of cuneiform and engraved clay tablets were discovered. Over the last 150 years, thousands have been translated, with more than one hundred thousand tablets to go! They reveal details of human life eight thousand years ago, an they "reintroduced" humans to the gods of old. They detailed the heavenly exploits of the gods and the earthly exploits of the humans.

3. Sitchin is the foremost interpreter of the Sumerian clay tablets. He elucidates historical details and connects them to other fields of knowledge. His

work stands up to comparative review with the work of other scholars. The following summary draws upon his books.

4. In the discovered Sumerian clay tablets, there exists *in our solar system* a twelfth planet called Nibiru. Now, I realize that the name Nibiru sounds weird and Star Trek—like but bear with me … Astronomers know it, validate it, and refer to it as Planet X. It is referred to as a planet, not a moon or an asteroid. This is a very important point to make. Sitchin writes that Nibiru has an *elongated orbit* which takes it far beyond Pluto, and that's why we can't *see* it. It passes near the sun and our planets *every 3,600 years*.

5. It is inhabited by race of ETs called the Anunnaki in the Bible as *Anakim*, which is referenced over 60 times. (Coincidence?) They are referred in cuneiforms scripts written on ancient Sumerian clay tablets as "those who from Heaven to Earth came." The tablets date back to 2150 BC. During one of its elliptical cycles of 3,600 years, Nibiru came into our cyclical orbit and was captured by our gravitational field, which did considerable damage to our original solar system. During one of its transits near the sun, around 500,000 years ago, the Anunnaki explored earth for precious metals that would shield them from a loss of atmosphere, based upon the gravitational force damage that occured. This precious metal was *gold*.

6. The following information which I'm about to describe may sound comical to some, but I urge you, the reader, to read with an open mind. Information comes directly from the Sumerian tablets. Is not "made-up material," mythology, or old wives' tales for children. It predates the Old Testament. It is, in my opinion, the *true* story of our origin. It is for those that have the energy to really test their belief system, think outside the box, and judge more on the rationality of this information rather than putting it into a science fiction category. It is meant for those that are gutsy enough to look inside themselves and really question what they have been told/sold. Ask yourselves, if this is true, does it make sense?

    Here we go:

7. The Anunnaki established a mining colony on earth in the Tigris-Euphrates Valley (near the modern city of Basra) which the Anunnaki called the Abzu. The expedition was supervised by Anu, who assigned his two sons, Enlil and Enki, to manage separate parts of the operation. Enlil was in charge of the area called E.DIN (Eden--coincidence?). Enki was responsible for Abzu in South Africa. Ruins as old as *100,000 years* have been found in current gold mines in South Africa.

    Enlil is considered God of Earth, as he was the ruler there, even over his brother Enki.

Enki would be known as the "snake" or "serpent" (Satan?) who tempted Eve. He was also the creator of humans.

Marduk became the warrior-god of Egypt and the Babylonians and built the pyramids.

Inanna became the primary Hindu deity and was considered the "mother" of the humans.

8. Throughout the course of development and the use of atomic weapons on Nibiru, the atmosphere became destabilized and began to die.

9. In order to protect Nibiru from atmospheric destruction, it was determined that a precious metal which could refract both heat and electricity that did not exist on Nibiru or any other neighboring planet—gold--was needed to be mined and exploded into the atmosphere to act as a "shield" from radiation (hence our current elevated economic value of gold).

10. An expedition was sent over 400,000 years ago to earth to mine for gold in the Tigris-Euphrates Valley.

11. As the quantities of gold became depleted from the Middle East to Africa and the hardships of mining operations increased , the Anunnaki team of miners (approximately 300 ETs) revolted and prompted their leaders to create a slave race "to do the heavy lifting."

12. That slave race was genetically developed from existing primates that already inhabited the earth, Cro-Magnons, into larger-brained *Homo Sapiens* with shorter life spans and greater intellect.

13. These *Homo Sapiens* (with a fused 45th and 46th chromosome) entered the Fertile Crescent and Mesopotamia (modern day Iraq) about 300,000 years ago.

14. Some of the Anunnaki interbred (evidence in the Bible) with these "favorable" *Homo Sapiens*.

15. Their offspring are what's referred to in the Bible as Nephilim (Genesis 6:4—"Nephilim … are the products of sexual union of heavenly beings … and human women") , The early humans were called Adamu, hence the name "Adam." Biblical texts state how the gods walked and talked with humans and had "intercourse with the daughters of man."

16. After the "cataclysm" which may either refer to a nuclear blast or a flood, the Anunnaki, who would later return to the earth, were called "fallen angels."

Genealogy of Later Mesopotamian Gods

Sumerian names given first, Akkadian last

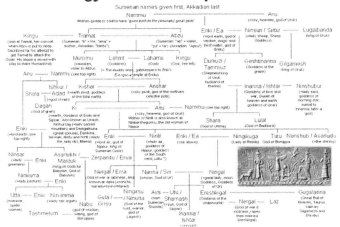

Nammu ──────────────────────────────── Anu
Mother-goddess said to have "given birth to the (Anunaki) great gods"   (=sky, heavens, god of Uruk)

Kingu   Tiamat ──── Abzu   Enki / Ea   Ninsun / Sirtur   Lugalbanda
(son of Tiamat, her consort   (Sumerian: "ti" = life, "ama" =   (Sumerian: "ab" = water,   (=lord earth, god of   (=lady sheep, Sheep   (king of Uruk)
when Abzu is put to sleep.   mother, Akkadian, "Tiamtu")   "zu" = far) (Akkadian, "Apsu")   wisdom, magic and   Goddess)
Sacrificed for his attempt to   freshwater, god of
get Tiamat to attack the   Enki)
Gods. His blood is mixed with   Mummu   Lahmu ──── Lahamu   Kingu   Dumuzi /   Geshtinanna
clay to make Humankind)   (=awaken,   (Goddess)   (God)   (God, see left)   Tammuz   (=Goddess of the   Gilgamesh
                                 vizier to Abzu)   (= The muddy ones), gatekeepers to Enki's.   (Shepherd king   grape)   (king of Uruk)
      Anu ──── Nammu (see top right)   Ea-engurra temple at Eridu)   of Uruk,
                                                                          husband of
         Ishkur /   Kishar ──────────── Anshar   Inanna):   Inanna / Ishtar   Ninshubur
    Shala ── Adad (=earth pivot, goddess   (=sky pivot, god of the northern   (Goddess of love and   (=lady east,
            (=god of rain)   of the total earth)   celestial pole)   war, Queen of   goddess of
      Dagan   heaven and earth,   morning star,
      (God of grain)   Ki ──────────── Anu ──── Nammu (see top right)   goddess of Uruk)   sukkal to
                                (=sky, heavens, god of Uruk)   Shara   Lulal   Inanna, later a
   Enki   Mother of Ninlil is also known at   (God of Umma)   (God of Badtibira)   god)
   (=lord earth, see   Ninharthegunnu, the old woman of
   top right)   Ninlil ──── Enlil ──── Ninlil   Enki / Ea ──────── Ningirsu   Tutu Nanshub / Asarludu
            (Ninmah, Nintu and Ninhi (=lady   (=lord air, god of   (=lady air,   (see above)   (Lady of Reeds)   God of Borsippa   (=the shining)
            (great spouse), Damkina,   Nippur, king of   goddess of
            Ninmah, Nintu and Ninhi (=lady   Sumerian Gods)   Nippur, goddess
            life, lady rib), Nlama)   of the South
   Ninsar ──── Enki   Asarluhhi /   wind, Lillu ")
   (=lady   Marduk ──── Zerpanitu / Erua
   greenery)   (King of Gods for
   Ninkurra ──── Enki   Babylon, God of   Nergal / Erra   Nanna / Sin   Ningal
   (=lady pasture)   Babylon)   (God of war or sickness, also   (=moon, God of Ur)   (=great lady, moon
   Utta ──── Enki ── Nin-imma   known as Apia (=nonhis-   Goddess, Goddess
   (=weaver,   (=lady (sic) organ)   Nabu   Girsu   Ningirsu   Aya ── Utu /   of Ur)   Gugalanna
   spider   god)   Gula /── Ninurta   (from   Shamash   Ereshkigal ──── Nergal ── Laz   (Great Bull of
   woman)   Tashmetum   (god of wisdom,   (God of the   Sumerian   (=Uruk, God of   (Goddess of the   (God of war &   Heaven, Tiamat,
                      writing, god of   hunt, warrior   Sheklla)   Sippar)   Underworld)   sickness, rapes   slain by
                      Borsippa)   god of   Inanna /   then marries   Gilgamesh and
                             Lagash)   Ishtar   Ereshkigal)   Enkidu)
                                                   (see right)

An Anunnaki God on Horseback

Sumerian Relief of the solar system and a cigar shaped
UFO

EVIDENCE OF THE ANUNNAKI

Cuneiform script written on clay tablets in ancient Sumerian texts define the Anunnaki as "those who from heaven to earth came."

The Sumerian texts describe the Anunnaki as blonde (with occasional redheads) with pale skin and blue eyes. Most Sumerian drawings depict them as being very tall-- almost eight feet in height. The original Goliath was an Anunnaki, while David was a human. They are typically depicted in Sumerian art as being clothed and wearing some sort of headdresses, indicating that their bodies are less adapted to the heat of our sun than ours are. Their headdresses are what Egyptian gods copied and used after the Anunnaki left earth.
Sumer was where the majority of the Anunnaki lived and therefore showed the quickest rise in civilized growth on earth.

The Anunnaki came to the earth because they needed gold. They used slaves in southern Africa to mine it. Isn't it a coincidence that today gold is such a precious (and expensive) material? It is our most valuable medium of exchange. The reason is that we humans are taught to value it *as a result of* it's importance to the Anunnaki. The Anunnaki needed the gold because it is the most resistant and reflective metal of heat. Anunnaki apparently blew holes in their ozone layer (just as we humans are doing now) and needed the gold to explode into their atmosphere to shield them from their suns heat.

Taken directly from the tablets:

"In the atmosphere a breaching has occurred … Nibiru's air has thinner been made, the protective shield has been diminished. To use a metal, gold was its name. On Nibiru it was rare; within the Hammered Bracelet it was abundant" (the Hammered Bracelet refers to our asteroid belt).

"The Anunnaki solution was to disperse extremely tiny flakes of gold into their upper atmosphere to patch the holes needing repair. [Ironically, modern scientists contend that if we are ever forced to repair our own damaged ozone layer, tiny particles of gold shot into the upper atmosphere would be the best way to go about it. Such tiny particles have ideal insulating and reflecting properties and will stay aloft indefinitely.] Unfortunately for the Anunnaki, Nibiru did not contain enough gold to allow completion of what had to be a monumental repair job. [Remember, Nibiru is roughly 3X-5X the size of Earth]" (Pye, *Everything You Know Is Wrong*).

The Anunnaki gods living on earth were at odds with how much power and intelligence to permit humans to have. Some gods had decided to give humans knowledge for survival. Some gods wanted to keep them only as intelligent as to work in mines and understand their instructions. This caused immense tension among the gods, which I believe initially led to argument and conflict and then bloody battles and even nuclear war. Various gods took refuge in different, remote parts of the earth, such as the Americas, China, Australia, and India. The gods controlled humans, rewarding them when appropriate or punishing them if they disobeyed. This mirrors the relationship between God and man and the Old Testament, where the "wrath of god" is mentioned several times.

Enlil was the god who did not like humans. He was considered the god of "wrath." Was he the feared god of the Israelites? Since Enlil and Enki didn't get along and feuded often, it is my belief that Enlil "chose" a particular group of humans to obey him without contest—the Israelites (hence the Chosen Ones). After choosing this group to obey any and all of his rules (The Ten Commandments), he rescued them from slavery which had been under Marduk's control. This is why the Israelites stayed loyal to Enlil, and he became *the* God of Israel.

Sitchin believes that the Anunnaki gods departed Mesopotamia as early as 2000 BC because of toxic fallout from the atomic weapons that destroyed Sodom and Gomorrah. The gods seemed to have returned to their mother planet, Nibiru. In their absence, religions flourished and preserved the memory and history of these physical gods of the Bible.

Enki was portrayed as the "serpent" or "devil." Enki was really the "good guy," while Enlil was the "bad guy."

Enki was the genetic maker of Adam, and he loved his human experiment.

Abraham, the father of the Jewish, Christian, and Muslim faiths, was the human obedient leader of Enlil, who was used against Marduk's self-proclamation of being the "only god" to the humans. Abraham's nephew, Lot, was also obedient to Enlil.

Sumerian Kings List. Possibly the most important script ever discovered by humans. It lists 149 kings and rulers on planet Earth.

It lists 10 rulers pre-flood that reigned for 240,000 years! The Anunnaki Life span was tens of thousands of years.

It also outlines the Anunnaki arrival to Earth from their "space colony" from "Heaven," and cities Ninurta (see Sumerian Familial Chart) as the biblical "Yahweh" who destroyed Sodom and Gomorrah.

www.genesisflood.blog.com

Sumerian Relief – This relief is incredible. It depicts an
Anunnaki ruler seated (Anu?), and two other Anunnaki.
Take a look at the solar system shown. It has 12 planets,
The Sun, the Earth, the Moon, (recognized at the time as
a planet which formed from Tiamat), Mercury, Venus,
Mars, Jupiter, Saturn, Uranus, Neptune, Pluto, and of
course, the Anunnaki planet of Nibiru. It clearly shows
Nibiru in our solar system.

WHY GOLD?

- It is virtually indestructible.
- It is recyclable.
- It is immune to air, water, and oxygen.
- It will not rust or tarnish.
- It is the most electrically conductive (can convey electrical current in extreme temperatures.)
- It is the most ductile of all the metals.
- It is the most reflective of all the metals (ideal for heat and radiation reflection).
- It conducts heat and transfers it.

"Let gold from the waters be obtained, let for salvation on Nibiru be tested" (Sumerian Tablets).

The Vatican holds most of the gold in the world. It is also a common conspiracy theory that Jesuit US presidents have transferred almost all the gold stored in Fort Knox to the Vatican! Is there really any gold there anyway?

Where and When Were They Mined?

In South Africa, the following mines and their respective ages were discovered:

8000 BC—mines found
20,000-26,000 BC—further search uncovered mines

41,000 BC in Swaziland; then lastly, after bones were uncovered, mines dating back to 50,000 BC were found.

How can this be possible? Mines with human bones found that are carbon-dated at 50,000 BC?

The Anunnaki came to earth in groups of fifty, until a maximum of six hundred were used to mine the gold initially out of South Africa.
Today, South Africa contains many, abandoned gold mines that are quite deep. Some evidence from charcoal indicates that these mines are over 100,000 years old! Doesn't this, in and of itself, prove the fact that the Anunnaki *did in fact* come here? How else, or--a better question--*why* else would early humans have mined gold deep in the earth? Their only instinct would be food, warmth, and safety … not gold.

Enki, who acted as the main scientist to produce *Homo sapiens* from hominoids, use the gene identified as *MYH16* to be the main gene directly modified to create advanced beings. His purpose was to speed up the evolutionary process of intelligent life on earth. A verse taken from the Sumerian tablets quotes Enki as saying, "I will produce a lowly primitive. 'Man' shall be his name. I will create a primitive worker; He will be charged with the service of the gods, that they might have their ease."

The "Adam" was named because he was created of the "adama," or the earth's soil. Again, this speaks about Adam being a group as opposed to a single human being, and earth's soil refers to the clay pots that Annunaki used to store the sperm and egg mixture.

After the flood, the Anunnaki moved operations to Peru and Bolivia, which had high mountains way above sea level, could still exist with floods, and had gold. Obviously, higher-ground civilizations such as Machu Picchu and Tihuanaco (Lake Titicaca) were born of this need. Tihuanaco (like Baalbek in Lebanon) also had enormous multi-ton obelisks with gold hollowed out on the *inside* of the obelisk. Why the inside? Obviously the Anunnaki use these enormous structures, with gold inside, as some conductive antenna or rod. These obelisks may have been used for communications.

It is said that the Anunnaki would return every 3,600 years, when close to earth. Our primitive earth was called "Tiamat" in Sumerian scripts. Is it possible that Tiamat was struck by a moon from an intruder planet, which cracked off a part Tiamat which is today our asteroid belt?

The Sumerians recorded Nibiru's orbit around the sun to be every 3,600 years (another divisible by 6). This was called a "shar." Therefore, every 3,600 years, Nibiru would circle our sun and be very close to earth. Is this why Anunnaki visit earth every 3,600 years? Because of proximity to us?

"Sumerian tablets say that Enlil was the administrative ruler of Earth, the god who wanted to let mankind die in the Great Flood, the author of the Ten Commandments, a fair but firm, often harsh, controller of his followers [this is the god of the Jews and Christians.] Enki appears as the God who suggests creating mankind in the image of the gods, the serpent in the Garden of Eden helping mankind to procreate, the god who told Noah how to build the Ark and the principle scientist/geneticists of Earth. Nibiru, then would be 'heaven'" (McDowell).

Mining in Africa was the impetus for creating mankind. The biblical story of the creation of Adam is a short version of Sumerian records of this event. The Anunnaki miners in Africa faced hard work and unpleasant working conditions. In the first labor strike on earth, they mutinied. Enlil recommended punishing the miners, but Anu (his father) was more understanding of their plight. To lift the burden of this work from the Anunnaki, Enki volunteered that a primitive worker could be produced by crossbreeding the Anunnaki and the rugged ape-men at the time, *Homo erectus*. This plan was readily accepted, and the *Homo erectus* eggs were fertilized with Anunnaki sperm and placed in Anunnaki women for birth. After much experimentation, the process finally produced human workers that were collectively called "Adam."

In the early 1900s, British archaeologist started doing excavations in the ancient Sumerian city of Ur. Many of the artifacts and tablets spoke of beings called the Anunnaki and depicted these beings with wings (similar to angels in our modern Bible?). The Anunnaki apparently had the power of flight.

MANNA

Based on the Bible's account of Moses leading the Jews out of Egypt for 40 days and nights through the desert without food, God showered "manna" upon them to eat. It has always been thought that manna was a kind of bread. But no more. In 1904, Sir W.M. Flinders Petrie found tons of white powdered gold at the top of the Sinai mountain (on which Moses is believed to have received the Ten Commandments.) Although the analysis of this white powder has never been made public, this white gold may have been used to make manna. Was it used to consume? The white gold contains both iridium and rhodium, two consumable elements (found in carrots). So was white gold used as food?

GOD(S)

"Let *us* make man in *our* own image after *our* likeness" (Genesis 1:26). "And the Lord God said, 'Behold the man has become one of *us*, to know good and evil'" (Genesis 3:22). The Hebrew word for God is "Elohim" which literally translates to "*all* the Gods"—all, not *one*." Come, let *us* go down, and there confound their language" (Genesis 11:7). Who is "us."

How much proof does one need? The Sumerian tablets and the Old Testament both use these phrases that claim that our makers are *plural*. While I grew up believing in just one God, the Supreme being, and none other, I now believe that we still do have one Supreme being, which exists not in human form, which has created an ordered universe. The Anunnaki are "our" gods, because they created us. There are physical, imperfect flesh and blood humanoids. They love, they fight, they speak, they learn, just like us.

Genesis 1:26 (Hebrew version), implies that a group of advanced beings identified by the word "Elohim," basically said, "Let us make an earthling in our own image according to our likeness." Interestingly, in Chapter 4, "Elohim" is dropped altogether, and from then on, YHVH (pronounced "Yahweh") is used alone. This is very interesting. It denotes that the group then renounced all power to *one god*, which became the one god—Hashem—of the Hebrews, Enlil!

God descended to the tent of Moses in a cloud. Afterward, when the Israelites saw Moses, his face was radiant. Apparently, the face to face encounter with God was a radioactive event and bleached his skin. Moses was 80 years old at the time of the Exodus, which means he would have been born in 1526 BC, during the reign of Thutmose I. Moses wrote the books of the Laws of the Hebrews. "Jehovah was indeed a being of flesh and blood who flew in a craft that created fire, wind, and noise. The vehicle was used to transport Moses to the summit of Mount Sinai as stated in Exodus [as] on eagle's wings and brought you to myself" (Marrs).

1. Over 300,000 cuneiform tablets have been found to date. About 120,000 are in British Museums.

2. Sitchin deciphered only 2,000 of these.

3. The Sumerians depicted the gods as being of flesh and blood.

4. There were no humans on earth for the first 200,000 years the Anunnaki were here.

5. The *Epic of Gilgamesh* was earth's first written story-- on twelve cuneiform tablets.

6. On the eleventh tablet of the *Epic of Gilgamesh*, it is written that the Anunnaki provided wheat, sheep, and other food to the survivors of the Great Flood. Evidence of 9,000 year old domesticated wheat has been found on the slopes of Mount Ararat in Turkey.

7. The reason the Anunnaki built gold mines in the Americas is that their original mines in Africa and the Middle East were filled with mud after the Great Flood.

8. The reason the steps on the pyramids of the Americas are so steep is that their builders--the Anunnaki--were over 8 feet tall.

CHAPTER 15
TESLA

Edison's nemesis! Tesla was the inventor of alternating current as well as several other inventions. Tesla was the *real* genius, not Edison. Edison adhered strictly to *direct current*, which restricted the capacity of the current to direct sources. Tesla, on the other hand, developed *alternating current*, which would allow for "free energy" worldwide by using wireless technology! This was developed in the early 1900s. Tesla was backed by the powerful Westinghouse Electric Company, who paid him royalties in exchange for use of his patents. Some of his magnificent achievements were these:

- The development of the polyphase alternating current system
- sending 1,000,000 volts through his body in order to light up a bulb in his hands, without injury
- harnessing the power of Niagara Falls to light up the city of Buffalo
- discovering the first rotating magnetic field (which powers UFOs?)
- The development of the first electronic tube
- the development of the fluorescent bulb
- the development of the radio

Tesla was determined to use the earth's geomagnetic energy and tap into it.

Tesla admitted in the early 1950s that he was actively "communicating" with aliens. After the Roswell crash and other crashes in Arizona and New Mexico, he made these statements publicly, which shocked the world. Here was a revolutionary scientist that broke almost all barriers of physics and outwitted the giant Edison at his own game.

In 1958, there were rumors in the military that Tesla had a notebook which contain numerous extraterrestrial symbols, codes, and forms that corresponded to symbols that the military found on alien gadgets. Tesla's notebook was confiscated by the FBI.

He was found dead, penniless, in a hotel in New York City.

I mention Tesla because he was the human inventor of wireless energy. My discussions about the pyramids and their ability to transmit energy leads me to believe that Tesla had tapped into the secret of the Anunnaki ability to communicate across Earth and move massive stones to use as building materials through levitation.

CHAPTER 16
GRAVITY AND GEOMETRY

Einstein's theory of relativity says that nothing can travel faster than the speed of light. While Einstein was one of the--if not *the*--greatest physicist and minds of all time, I believe that this may not be entirely true. There is an "ether" particle described in Newton's interactions of gravity, electromagnetism, and light. Gravity is a *push* , not a *pull*. Tiny particles called gravitons are what push objects. Gravitons are one component of the invisible ether. They interact with the component of the ether that transmits light but are distinct from them. "Gravitons propagate the force of gravity much faster than the speed of light. In fact, observations of binary pulsars show that the speed of gravity must be at least 20 billion times the speed of light" (McDowell).

So how then did light waves travel through a vacuum of nothingness? The answer: the *ether*. Basically, the universe consist of mass (electrons, protons, neutrons, and so on) and electromagnetic radiation (light, radio waves, UV radiation, X-Rays , and so on). All of this is embedded in a space-time "plenum," , which came into existence from nothingness approximately 15 billion years ago.

In his studio, T. Townsend Brown, an American physicist in the 1930s, created saucers that had no propellers, jets, or moving parts. They created a gravitational field around themselves. They acted like a surfboard on a wave. The electro-gravitational saucer created its own "Hill," which is a local distortion of the gravitational field. Then it "takes" or "rides" this Hill with it in any chosen direction and at any rate. Whereas our turbojets attempt to combat the strength of gravity through brute force, electrogravitics instead attempts to control gravity and make it work *for it* instead of *against it.*

ANTI-GRAVITY

Anti-gravity is an important factor in discussing how the ETs travel. Since the Roswell crash, the Air Force has been experimenting with anti-gravity. In 1957, at least 14 US universities and research facilities were researching it Lawrence D. Bell, founder of Bell Aircraft, said, "We are already working with nuclear fuels and equipment to cancel out gravity instead of fighting it ... make no mistake about it, anti-gravity motors and G-ships are coming." Ben R. Rich, the ex-president of Skunk Works, Lockheed Martin's Advanced Development Programs Group, in 1993 at a speech at the University of California, said, "We already have the means to travel among the stars, but these technologies are locked up in black projects and it would take an act of God to ever get them out to benefit humanity. ... *Anything you can imagine, we already know how to do.*"

According to Dr. Richard Boylan, we now have anti-gravity craft capable of routine intra-solar system travel and deep-space ships referred to as "Nautilus types" that use sophisticated propulsion systems such as *tachyon* and *anti-matter engines* (see http://www.drboylan.com). He also claims that we have had military bases on the moon *and* Mars since 1962!

The B-2 stealth bomber, which was developed in the 1970s, shows us evidence of anti-gravity propulsion. Marion Williams, a former CIA agent who worked in Area 51 where the B-2 was test flown, said, "The design principles from crashed alien anti-gravity spacecraft were being utilized in the stealth bomber" (LaViolette).

Basically, electrogravitics is a technology that allows a spacecraft to artificially alter its own gravity field in such a manner that it is able to levitate itself. As early as 1958, a small scale model of an electrogravitic-powered aircraft was able to lift 110% of its own weight.

From 1952 to 1957, a project called Skyvault was initiated to develop an anti-gravity vehicle that used microwave beams as its means for propulsion. Why microwave beams? In 1957, during one aerial chase with a UFO near Meridian, Mississippi, a US-made RB-47 plane picked up a three-gigahertz signal of microwave frequency emanating from the UFO!

Project Redlight was code named for the recovery of captured UFOs and the development of anti-gravity propulsion from them. It is estimated that as many as 16 UFO craft have crashed at various locations around the world between 1947 and 1986.

A Dr. Richard Oboussy is experimenting with a form of 'warp drive' to make it possible to visit the stars. His 'In-Space Propulsion Technology Program' has already tested an effective ion-propulsion system and is currently working on the development of solar sail technology" (Davis).

EVIDENCE OF REVERSE ENGINEERING

- Patent #2,949,550—July 3, 1957—filed by the legendary father of anti-gravity technology, T. Townsend Brown, demonstrates the anti-gravity action of highly charged transducers.
- Patent #4,663,932—July 26, 1982—filed by James E. Cox, is called the "Dipolar Force Field Propulsion System," a type of magnetohydrodynamic thruster, believed to be a primary force of UFO propulsion. It requires no fuel but rather uses "a propellant comprised of neutral particles of matter having an electric dipole characteristic." It absorbs microwave energy and is invisible to radar.
- Patent #5,197,279—March 6, 1992—filed by James R. Taylor—is an engine that uses electromagnetic energy produced by supercooled, high-density electric power. It creates ripples on the fabric of space-time on which to ride a craft.
- Patent #5,269,482—September 30, 1991—filed by Ernest J. Shearing, it is for the "Protective Enclosure Apparatus for Magnetic Propulsion Space Vehicle." This was an invention made to protect the astronauts and occupants of a spacecraft from powerful electromagnetic and acceleration forces.
- Contract #33(038)-3736—1949—written by C.W. Simmons, C.T. Greenridge, and C.M. Craighead— titled "Second Progress Report Covering the Period 9/1/49-10/21/49 on Research and Development on

Titanium Alloys." This contract was released by the FOIA (Freedom of Information Act), was produced for Wright-Patterson Air Force Base (the destination of the Roswell craft and EBE) regarding "shape memory alloys" (like the ones found at Roswell). This alloy was so strong that a bullet fired from a gun could not penetrate it. It was so flexible that you could crumple it up like a Kleenex™ tissue. But the most interesting part is that after you clump it up, it returns to its original flat surface by itself (Kasten). All of the Above patents reflect the fact that we are well-ahead of our perceived position in technology. We were seeded this technology by ET's and have run with it.

LEVITATION

How were the pyramids *really* built? How could massive, precision-cut obelisks--like those found at the Baalbek-- have been lifted? These are hundreds, thousands of tons! Humans could not have lifted them, no matter how many humans tried. Humans could not have carved them with such precision. Even stonemasons today admit that they could not have cut the obelisks at the Baalbek with today's 21st century machines.

Impossible. So how was it done? I believe it was done by *levitation*. T. B. Pawlicki's book *How To Build A Flying Saucer* says, "I believe the way the ancients transported megaliths was to attach a small tuning fork to each stone, causing the module to levitate when the properly tuned vibration was sounded." In Tibet, monks have levitated stone through the sound of 6 long trumpets and 13 open ended drums pointing at the stone. Acoustic levitation. Can sound manipulate and/or nullify gravity?

Even in the United States there have been accounts of heavy stone levitation. In Homestead, Florida, there exists a remarkable stone construction called Coral Castle. It was built from 1923 to 1951 by Edward Leedskalnin, now deceased, a Latvian immigrant who worked alone and only late at night. The numerous structures are built from stone blocks averaging 6 tons (12,000 pounds), double the weight of an average great pyramid stone (6000 pounds). One stone block weighs 30 tons! He claimed that he discovered how the pyramids of Egypt were built. His workshop contains the following:

- bar magnets with alternating polarity: and
- a turning flywheel, both manual and by gasoline engine

The flywheel generated an alternating current. The current transmitted to the stone blocks as sound/magnetic waves. The magnetic waves set the gravitons in motion and moved the stone. Simple!

An amazing theory of levitation was described in the book *Ancient Technology in Peru and Bolivia* by the great author David Hatcher Childress. It involves quartz crystal and electricity. Here are the details:

When a crystal is struck with high voltages, [it] will "bend" … It will give off a piezoelectric signal and, incredibly, it actually loses the gravitational force that would naturally pull it toward the center of the mass [in this case, the earth]. The crystal then becomes essentially weightless, no matter how heavy it was before being bent by high voltages … Then gigantic blocks of granite, which are full of small quartz crystals, could theoretically be moved with very little effort, no matter what they weigh … it contracts or expands—it "bends" and flexes, changing its shape. This sudden change in the crystal causes the rectangular granite-crystal block to "jump." It also becomes briefly weightless during this period and could be "pushed" forward.

Canadian experimenter John Hutchinson has been experimenting with this technique for decades. You can view his experiments on You Tube.

Levitation through alternating current, magnets, and low sound…that's how they did it.

An interesting note about geometry and its relationship to the genesis of life is that the same numbers frequently come up as the "building blocks." Such numbers are three and six. In Masonic tradition, to achieve the level of a 33rd degree Mason is the ultimate goal. If you look at a tetrahedron—the shape of the pyramids—you will find that all shapes are a derivative of it.

Even our DNA spiral can be derived from tetrahedrons. According to *Pyramids of Montauk* author Preston Nichols says, "If you stack 33 [again that number] tetrahedrons face to face, it will form a spiral or helix shape. If you splice the spiral of 33 tetrahedrons in half, it will form two helixes which duplicates the form of DNA exactly. Once this is done, the number of solid geometric faces on the two helixes is 66.666. All life on this planet is considered to be carbon based because the carbon atom is conspicuously present in all living things. Carbon has 6 electrons, 6 protons and 6 neutrons."

Also it's interesting that the Bible listed the human age maximum of "120 years" another multiple of 6.

Levitating Top

Here are some interesting facts concerning the number 33:

- The first atomic bomb was detonated at the Trinity site at White Sands Proving Grounds near Socorro, New Mexico at 33 degrees latitude.
- JFK was assassinated in Dallas (near the Trinity River!) At 33 degrees latitude.
- The Watchers descended to earth on Mount Hermon, which lies on the 33rd parallel, which is a latitude of 33 degrees north of the equator. Interestingly, if you trace the 33rd parallel to the exact geographic global opposite from Mount Hermon, you will be at Roswell, New Mexico! (Coincidence?) They are exact polar opposites on the same 33 degree north latitude (Roberts).

CHAPTER 17
WORLD GRID

Time is what keeps everything from happening at once.

--Ray Cummings, science fiction writer, 1922

Why did ETs build pyramids where they did? Why build Stonehenge in the middle of nowhere in England? Why did the Nazis build a headquarters in a wooded swamp? The answer is that these locations fall directly on a grid system on earth that harnesses energy. The ETs knew this. If you observe the earth from the sky and map out all the pyramids and monuments built on the earth, you will notice that they fall on a geometric grid--almost like slicing up the earth's landmass into zones that all fall along a line.

I personally despise everything about the Nazis and their brutal and disgusting forms of torture and slaying of Jews and Christians alike. They were obsessed with ETs , UFOs, and harnessing any power known that would give them an advantage in World War II. A "Thule Society" was created to harness what they called the "Vril," or mysterious geomagnetic power drawn from the earth through its grid.

"Ancient sites, temples, and structures are laid out upon a grid whose significance is determined by their placement at mathematically significant places upon the surface of the Earth, or in correspondence with celestial, astronomical, alignments, or both, and often incorporate these and other mathematical analogues in the structures themselves" (Farrell, *The Grid of the Gods*).

All of the following pyramids—Angkor Wat, Teotihuacan, Tikal, Stonehenge, and Avebury—all fall along the grid.

Pyramid counts—The following list the amount of pyramids that exist per country:

- Egypt—85
- Sudan—100+
- China—100+
- Mexico—200+
- Guatemala—30
- Ecuador—10
- Peru—200+
- Bolivia—1
- Tibet—1
- Bosnia—1

(Carl P. Munck, Whispers from time: The Pyramid Bible)

Monuments—10,000+ worldwide

THE PHILADELPHIA EXPERIMENT

While this experiment orchestrated by the US government is not directly related to aliens or evolution, it is a startling discovery in and of itself. Scientist Einstein and Tesla were assigned to make a US Battleship *disappear*! The terminology used was "degaussing". It was used as a military tactic for tricking the enemy. Several books were written--and a movie was made--about this experiment. It took place in 1943 with the USS *Eldridge* destroyer in the waters off of the Philadelphia (hence the name) Naval Shipyard.

Essentially, they generated an enormously powerful electromagnetic field, mostly in the range of radio waves, around the ship. This intense field was powered by harnessing the vast amounts of telluric "free energy" emanating out of the earth itself. This energy field is referred to as the "geomagnetic grid," as it manifests itself in intersecting lines around the earth, called "ley lines" by students of European megalithic sites.

The experiment was known as Project Rainbow, and its objective was to create complete invisibility of the carrier and teleport it to another location. According to Alexandra Bruce, author of *The Philadelphia Experiment Murder*, the war ship disappeared for about 20 minutes and was sighted in many different places. Different accounts had it appearing as "a giant quantum particle" off the coast of Norfolk Virginia, Northern Italy, and Montauk Point, New York (three grid points?). When the *Eldridge* finally reappeared in Philadelphia, several men aboard the ship were dead, and some were *melded into the steel of the ship*! Some were "phasing" in and out of view, and some went insane from what they reported was a foray into hyperspace.

The U.S.S. Eldridge

Aboard the U.S.S. Eldridge. "Val Thor," the blonde circled is reputedly a "Nordic Alien" who helped the U.S. Government with the experiment. (I cannot validate this point).

Proof of the Philadelphia Experiment. – This letter dated 4/23/1953, ten years after the event, mentions "DE-173" (The U.S.S. Eldridge), and that seven members of the original crew escaped from their "Psychol Unit in Virginia" and must be "captured." It also verifies that Nicola Tesla's "Spatial Analyzer." This memo was written by the captain of the U.S. Navy to Edwin Condon who wrote "The Condon Report" in the 1960s debunking UFOs.

CHAPTER 18
TIME TRAVEL

Summers, when I was younger, I would go to the beaches of East Hampton, NY to swim and enjoy the sun. Little did I know that every time I drove out of there, I passed one of the most famous time-travel experimentation facilities in the world--Montauk--right past East Hampton and Amagansett, right on the shore at Camp Hero. The Montauk project was a government-sponsored experiment in time travel. Brookhaven Labs, farther east on Long Island, did the R&D, and Camp Hero put it into practice. I never believed in time travel before, but after reading several books on the topic by people like Preston Nichols and Al Bielek, I wasn't so sure anymore.

This area was chosen because it houses a huge SAGE radar antenna that admitted a frequency of 400--425 megahertz, coincidentally the same band used to enter the consciousness of the human mind. There is also a circular, stone particle beam accelerator. It is also thought that the area is a part of the earth's geomagnetic "grid" or falls on a line (like a tectonic plate) of geometric shapes that crisscross the earth and carry electromagnetic force. According to Al Bielek (crewmember of the USS *Eldridge*), the earth's magnetic energy peaks every 20 years on the 12th of August. The Philadelphia experiment was reportedly conducted on

August 12, 1943. It allowed a synchronicity of electromagnetic events to occur which purportedly rendered the invisibility of a battleship!

As discussed in the last chapter, some of the greatest minds including Einstein, Tesla, and Von Neumann, were a part of this experiment, which was initially developed to cloak ships for military defensive purposes, but the process of creating an invisibility shield supposedly fused crew members to the hull of the ship! The crew members would "blink out" for periods of time. K.B. Wells, author of *The Montauk Files: Unearthing the Phoenix Conspiracy*, attributes this to a "misalignment of their corporeal time-center. "

Follow-up study was made at the Brookhaven Center (which still exists today) on Long Island, New York. All experiments halted, of course, but a great deal was learned about the use of electromagnetic analog and timed pulsing focused on an object during this period.

Al Bielek, a survivor of the experiment, also quoted, "There are five dimensions to our reality: our 3-D time [which includes for dimensions] and 'T2' which vector was being rotated in these [The Philadelphia Experiment] to give time/space shifts" (Childress, *The Time Travel Handbook*).

CHAPTER 19
THE WINGMAKERS

The following is a true story (see Childress).

In 1972, in a remote section of northern New Mexico, a group of hikers discovered unusual artifacts and pictographs in an obscure canyon. An archaeologist from the University of New Mexico analyzed the artifacts and searched the area. All but one of the artifacts were dated to the eighth century AD. The one that wasn't was indecipherable. It appeared to be a compass of some sort but was made from an unknown technology. The compass was covered in strange hieroglyphic symbols, some of which were found on pottery also discovered. The pictographs were unlike any found in the American Southwest from any ancient tribes from Mesoamerica. Because this compass was so strange and the pictographs were indecipherable, the entire project was handed over to the US government's National Security Agency. It was code-named "Ancient Arrow." Initial expeditions of the site uncovered more pictographic symbols but nothing else.

However, after a series of rockslides in 1994, a hidden cavern was exposed within the canyon walls. At the back of the cavern, the research team discovered a well-hidden entrance into the interior of the canyon wall or rock structure of the Ancient Arrow site. There they found a system of tunnels and chambers that had been carved out from solid rock. There were a total of twenty-three chambers, all intricately connected to an interior corridor, and each chamber held a specific wall painting, series of pictographs, written hieroglyphs, and what seemed to be dormant alien technology!

The project was formally brought under the jurisdiction of the Advanced Contact Intelligence Organization (ACIO), which organized an interdisciplinary research team to discover the purpose of this extraterrestrial site. The ACIO is a secret or unacknowledged department of the NSA, headquartered in Virginia. It is largely unknown, even to senior members of the NSA. Its agenda is to research, assimilate, and replicate any technologies or discoveries of extraterrestrial origin. The members of the ACIO are brilliant thinkers, linguists, and hieroglyphic historians. After seven months of restoration, cataloguing, and analysis, a hypothesis was developed that an extra-terrestrial culture established an earth colony in the eighth century and isolated itself within the Ancient Arrow (USA) canyon. Their mission was to leave behind a massive 'time capsule' that would prove to be discovered in the late 20th century. It appeared to be a "cultural exchange" of sorts.

Years of research uncovered the following:

- Twenty-three separate but linked chambers formed together purposefully
- an optical disk discovered in the twenty-third chamber

Scientists analyzed the disk for one year but could not figure out how to access its content. It was placed in storage for another year. One day in the summer of 1996, a linguistics expert had an insight on how to unlock it--by mimicking the symbols of the wall paintings, *to symbols found in ancient Sumerian text*s! By placing the twenty-three words in the same order as the Ancient Arrow chambers, he was finally able to unlock the disk!

"The connection between the Sumerian language and the time capsule was the breakthrough the ACIO team had been waiting for. A simple set of twenty-three words elicited over 8000 pages of data from the optical disk. Unfortunately, the data was incomprehensible because there was no character set in the computer that could emulate the hieroglyphics and unusual symbols of the language. Thus, a translation index needed to be developed, which took an additional six months" (Childress, *Time Travel*).

As the chambers started to be translated, it was shown that each unit contained philosophy, science, poetry, art, music, and an introduction to the culture and identity of its creators. The creators refer to themselves as "WingMakers." They live 750 years *in our current future*! They claim to have left seven time capsules in various parts of the world.

In 1997, one ACIO scientist who worked on the project defected. He became a spokesperson for the public because of the importance of this discovery. His name has not been revealed to the public , but he has been interviewed by authors of papers written on the subject of this find. Some excerpts of his interview are as follows:

- A team of seven other scientists worked on the project from the ACIO and researched and catalogued all of the artifacts in all of the twenty-three chambers.
- An optical disk was found in the twenty-third chamber.
- The translation of the 8,110 pages of text found within the disc was called "granularity."
- The site was visited by the WingMakers in the eighth century.
- The WingMakers claim that the three-dimensional, five-sensory domain that we humans have adjusted to is the reason we are only using a fractional portion (10%) of our intelligence. They claim that the time capsule would be the bridge from the three-dimensional, five-sensory domain to the multidimensional seven-sensory domain.
- Earth is considered a very special planet because of its diverse body of ecosystems and natural resources. It is attractive to other species of life.

This is a spectacular story! Humans from 750 years in the future come back to earth to drop seven time capsules here, now, so that we can bridge the multidimensional and sensory gap and expand our intelligence!

CHAPTER 20
HUMAN CONTROL
OVER REAL HISTORY

Give me control over a nation's currency, and I care not who makes its laws.

--Mayer Amschel Rothschild (1743-1812)

There is a power organized, so subtle, so complete, so pervasive, that they had better not speak above their breath when they speak in condemnation of it.

President Woodrow Wilson

Three hundred men, all of whom know one another, direct the economic destiny of Europe and choose their successors from among themselves.

Walter Rathenau, General Electric, 1909

The USA is not run by its would-be "democratic government." Nothing could be more pathetic than the role that has to be played by the President of the United States, whose power is approximately zero.

R. Buckminster Fuller

The media may not always be able to tell us what to think, but they are strikingly successful in telling us what to think about.

Michael Parenti

War is a racket … War is largely a matter of money. Bankers lend money to foreign countries and when they cannot pay, the President sends Marines to get it.

Marine Major General Smedley Butler (1881-1940)

The Bilderberg Group

The Bilderberg Group is the most powerful secret organization in the world, organized in 1952 at Hotel Bilderberg in the Netherlands. Prince Bernhard of the Netherlands is so powerful that he can veto the Vatican's choice of the pope! His family, the Hapsburgs, are direct descendants of Roman Emperors, who are direct descendants of the House of David. The Bilderberg Group has been waging a "quiet war" on society. They control world banks, world politicians, world economies, and the top world military leaders. Their intent is complete world domination through a totalitarian socialist/feudalistic state controlled by approximately three hundred persons. These three hundred persons originate in all major countries of the world with direct routes to rulers, emperors, kings, and presidents. They initiate war and then lend money to the countries that fight. They fluctuate currencies and then lend money to those both in need and those that want to speculate. They insert specifically predetermined leaders as presidents of countries by any means necessary. They create currencies that note "Novus Ordo Seclorum" which translates to "New World Order." They have a penchant for the number thirteen, which is considered by layman as an "unlucky number." Why? Did you ever

consider why some buildings have no thirteenth floor?
Consider the following on the one-dollar bill:

- 13 leaves in the olive branches
- 13 bars and stripes
- 13 arrows
- 13 letters in "E Pluribus Unum"
- 13 letters in "Annuit Coeptis"
- 13 stars in the green crest
- 13 stines in the pyramid

Consider also the number thirty-three:

- A 33rd degree Mason is the highest plateau a mason can achieve in life.
- Because of gravity, when you pile salt on the table, it forms a conical mound with a slope of 33 degrees.
- Gravity's speed is 33 feet per second.
- In Qabalah, one strives to reach the 33rd principle of enlightenment
- DNA strands are 33 degrees
- Water turns to ice at 32.72 degrees Fahrenheit.
- Water contains 33 different substances.
- If you stack 33 tetrahedrons face to face, it forms a spiral or helix (DNA) shape.
- 66 (the number aligned in modern times with the devil or evil) is the number of the two DNA helixes.

Coincidence?

My point in discussing the Bilderbergs is to demonstrate suppression. The Bilderberg group is an all-powerful elite that suppresses any information that threatens their global agenda. The topics in this book, the knowledge, the information, is *threatening* to the Bilderbergs. If there exists an alien species that originated human life and that is *more powerful* than they are, there is a threat to their existence, power, and wealth.

This unclassified copy is for research purposes. Rev X; 2/26/56; Halvatica Insiders; Times text

RESTRICTED

SOM1-01

TO 12D1—3—11—1
MAJESTIC—12 GROUP SPECIAL OPERATIONS MANUAL

EXTRATERRESTRIAL
ENTITIES AND TECHNOLOGY,
RECOVERY AND DISPOSAL

TOP SECRET/MAJIC
EYES ONLY

WARNING! This is a TOP SECRET—MAJIC EYES ONLY document containing compartmentalized information essential to the national security of the United States. EYES ONLY ACCESS to the material herein is strictly limited to personnel possessing MAJIC—12 CLEARANCE LEVEL. Examination or use by unauthorized personnel is strictly forbidden and is punishable by federal law.

MAJESTIC—12 GROUP • *APRIL 1954*

MJ—12 4838BMAN 270435¹-54-1

215

This unclassified copy is for research purposes. Rev 3; 2/28/95; Helvetica headers, Times text

TOP SECRET/MAJIC EYES ONLY

SOM—01

Special Operations Manual
No. 1-01

MAJESTIC—12 GROUP
Washington 25, D.C., *7 April 1954*

EXTRATERRESTRIAL ENTITIES AND TECHNOLOGY RECOVERY AND DISPOSAL

TOP SECRET/MAJIC EYES ONLY
REPRODUCTION IN ANY FORM IS FORBIDDEN BY FEDERAL LAW

٦216٦

This unclassified copy is for research purposes. Rev 3; 2/28/95; Helvetica headers, Times text

TOP SECRET/MAJIC EYES ONLY

CHAPTER 1

OPERATION MAJESTIC—12

Section I. PROJECT PURPOSE AND GOALS

1. Scope

This manual has been prepared especially for Majestic—12 units. Its purpose is to present all aspects of Majestic—12 so authorized personnel will have a better understanding of the goals of the Group, be able to more expertly deal with Unidentified Flying Objects, Extraterrestrial Technology and Entities, and increase the efficiency of future operations.

2. General

MJ—12 takes the subject of the UFOBs, Extraterrestrial Technology and Extraterrestrial Biological Entities very seriously and considers the entire subject to be a matter of the very highest national security. For that reason everything relating to the subject has been assigned the very highest security classification. Three main points will be covered in this section.

    a. The general aspects of MJ—12 to clear up any misconceptions that anyone may have.

    b. The importance of the operation.

    c. The need for absolute secrecy in all phases of operation.

3. Security Classification

All information relating to MJ—12 has been classified MAJIC EYES ONLY and carries a security level 2 points above that of Top Secret. The reason for this has to do with the consequences that may arise not only from the impact upon the public should the existence of such matters become general knowledge, but also the danger of having such advanced technology as has been recovered by the Air Force fall into the hands of unfriendly foreign powers. No information is released to the public press and the official government position is that no special group such as MJ—12 exists.

4. History of the Group

Operation Majestic-12 was established by special classified presidential order on 24 September 1947 at the recommendation of Secretary of Defense James V. Forrestal and Dr. Vannevar Bush, Chairman of the Joint Research and Development Board. Operations are carried out under a Top Secret Research and Development—Intelligence Group directly responsible only to the President of the United States. The goals of the MJ-12 Group are as follows:

    a. The recovery for scientific study of all materials and devices of a foreign or extraterrestrial manufacture that may become available. Such material and devices will be recovered by any and all means deemed necessary by the Group.

    b. The recovery for scientific study of all entities and remains of entities not of terrestrial origin which may become available though independent action by those entities or by misfortune or military action.

    c. The establishment and administration of Special Teams to accomplish the above

MJ—12 4838B   2

TOP SECRET/MAJIC EYES ONLY
REPRODUCTION IN ANY FORM IS FORBIDDEN BY FEDERAL LAW

TOP SECRET/MAJIC EYES ONLY

operations.

d. The establishment and administration of special secure facilities located at secret locations within the continental borders of the United States for the receiving, processing, analysis, and scientific study of any and all material and entities classified as being of extraterrestrial origin by the Group of the Special Teams.

e. Establishment and administration of covert operation to be carried out in concert with Central Intelligence to effect the recovery for the United States of extraterrestrial technology and entities which may come down inside the territory of or fall into the possession of foreign powers.

f. The establishment and maintenance of absolute top secrecy concerning all the above operations.

5. Current Situation

It is considered as far as the current situation is concerned, that there are few indications that these objects and their builders pose a direct threat to the security of the United States, despite the uncertainty as to their ultimate motives in coming here. Certainly the technology possessed by these beings far surpasses anything known to modern science, yet their presence here seems to be benign, and they seem to be avoiding contact with our species, at least for the present. Several dead entities have been recovered along with a substantial amount of wreckage and devices from downed craft, all of which are now under study at various locations. No attempt has been made by extraterrestrial entities either to contact authorities or to recover their dead counterparts of the downed craft, even though one of the crashes was the result of direct military action. The greatest threat at this time arises from the acquisition and study of such advanced technology by foreign powers unfriendly to the United States. It is for this reason that the recovery and study of this type of material by the United States has been given such a high priority.

TOP SECRET/MAJIC EYES ONLY

This unclassified copy is for research purposes. Rev 3; 2/28/95; Helvetica headers, Times text

CHAPTER 2

INTRODUCTION

Section I. GENERAL

6.   Scope

a. This operations manual is published for the information and guidance of all concerned. It contains information on determination, documentation, collection, and disposition of debris, devices, craft, and occupants of such craft as defined as Extraterrestrial Technology or Extraterrestrial Biological Entities (EBEs) in Section II of this chapter.

b. Appendix I contains a list of current references, including technical manuals and other available publications applicable to these operations.

c. Appendix II contains a list of personnel who comprise the Majestic-12 Group.

7.   Forms and Records.

Forms used for reporting operation are listed in Appendix Ia.

Section II. DEFINITION AND DATA

8.   General

Extraterrestrial Technology is defined as follows:

a. Aircraft identified as not manufactured in the United States or any terrestrial foreign powers, including experimental military or civilian aircraft. Aircraft in this category are generally known as Unidentified Flying Objects, or UFOBs. Such aircraft may appear as one of several shapes and configurations and exhibit extraordinary flight characteristics.

b. Objects and devices of unknown origin or function, manufactured by processes or of materials not consistent with current technology or scientific knowledge.

c. Wreckage of any aircraft thought to be of extraterrestrial manufacture or origin. Such wreckage may be the results of accidents or military action.

d. Materials that exhibit unusual or extraordinary characteristics not consistent with current technology or scientific knowledge.

Extraterrestrial Biological Entities (EBEs) are described as:

a. Creatures, humanoid or otherwise, whose evolutionary processes responsible for their development are demonstrably different from those postulated or observed in homo sapiens.

9.   Description of Craft

Documented extraterrestrial craft (UFOBs) are classified in one of four categories based on general shape, as follows:

a. Elliptical, or disc shape. This type of craft is of a metallic construction and dull aluminum in color. They have the appearance of two pie-pans or shallow dishes pressed together and may have a raised dome on the top or bottom. No seams or joints are visible on the surface, giving the impression of one-piece construction. Discs are estimated

MJ—12 4838B

This unclassified copy is for research purposes. Rev 3; 2/28/95; Helvetica headers, Times text

TOP SECRET/MAJIC EYES ONLY

from 50-300 feet in diameter and the thickness is approximately 15 per cent of the diameter, not including the dome, which is 30 per cent of the disc diameter and extends another 4-6 feet above the main body of the disc. The dome may or may not include windows or ports, and ports are present around the lower rim of the disc in some instances. Most disc-shaped craft are equipped with lights on the top and bottom, and also around the rim. These lights are not visible when the craft is at rest or not functioning. There are generally no visible antenna or projections. Landing gear consists of three extendible legs ending in circular landing pads. When fully extended this landing gear supports the main body 2-3 feet above the surface at the lowest point. A rectangular hatch is located along the equator or on the lower surface of the disk.

b. Fuselage or cigar shape. Documented reports of this type of craft are extremely rare. Air Force radar reports indicate they are approximately 2 thousand feet long and 95 feet thick, and apparently they do not operate in the lower atmosphere. Very little information is available on the performance of these craft, but rader reports have indicated speeds in excess of 7,000 miles per hour. They do not appear to engage in the violent and erratic maneuvers associated with the smaller types.

c. Ovoid or circular shape. This type of craft is described as being shaped like an ice cream cone, being rounded at the large end and tapering to a near-point at the other end. They are approximately 30-40 feet long and the thick end diameter is approximately 20 per cent of the length. There is an extremely bright light at the pointed end, and this craft usually travels point down. They can appear to be any shape from round to cylindrical, depending upon the angle of observation. Often sightings of this type of craft are elliptical craft seen at an inclined angle or edge-on.

d. Airfoil or triangular shape. This craft is believed to be new technology due to the rarity and recency of the observations. Radar indicated an isosceles triangle profile, the longest side being nearly 300 feet in length. Little is known about the performance of these craft due to the rarity of good sightings, but they are believed capable of high speeds and abrupt maneuvers similar to or exceeding the performance attributed to types "a" and "c".

10. Description of Extraterrestrial Biological Entities (EBEs)

Examination of remains recovered from wreckage of UFOBs indicates that Extraterrestrial Biological Entities may be classified into two distinct categories as follows:

a. EBE Type I. These entities are humanoid and might be mistaken for human beings of the Oriental race if seen from a distance. They are bi-pedal, 5-5 feet 4 inches in height and weigh 80-100 pounds. Proportionally they are similar to humans, although the cranium is somewhat larger and more rounded. The skin is a pale, chalky-yellow in color, thick, and slightly pebbled in appearance. The eyes are small, wide-set, almond-shaped, with brownish-black irises with very large pupils. The whites of the eyes are not like that of humans, but have a pale gray cast. The ears are small and not low on the skull. The nose is thin and long, and the mouth is wider than in humans, and nearly lipless. There is no apparent facial hair and very little body hair, that being very fine and confined to the underarm and the groin area. The body is thin and without apparent body fat, but the muscles are well-developed. The hands are small, with four long digits but no opposable thumb. The outside digit is jointed in a manner as to be nearly opposable, and there is no webbing between the finger as in humans. The legs are slightly but noticeably bowed, and the feet are somewhat splayed and proportionally large.

MJ—12 4838B   5

ADAM = ALIEN

TOP SECRET/MAJIC EYES ONLY

b. Type II. These entities are humanoid but differ from Type I in many respects. They are bi-pedal, 3 feet 5 inches-4 feet 2 inches in height and weigh 25-50 pounds. Proportionally, the head is much larger than humans or Type I EBEs, the cranium being much larger and elongated. The eyes are very large, slanted, and nearly wrap around the side of the skull. They are black with no whites showing. There is no noticeable brow ridge, and the skull has a slight peak that runs over the crown. The nose consists of two small slits which sit high above the slit-like mouth. There are no external ears. The skin is a pale bluish-gray color, being somewhat darker on the back of the creature, and is very smooth and fine-celled. There is no hair on either the face or the body, and these creatures do not appear to be mammalian. The arms are long in proportion to the legs, and the hands have three long, tapering fingers and a thumb which is nearly as long as the fingers. The second finger is thicker than the others, but not as long as the index finger. The feet are small and narrow, and four toes are joined together with a membrane.

It is not definitely known where either type of creature originated, but it seems certain that they did not evolve on earth. It is further evident, although not certain, that they may have originated on two different planets.

11. Description of Extraterrestrial Technology

The following information is from preliminary analysis reports of wreckage collected from crash sites of extraterrestrial craft 1947-1953, excerpts from which are quoted verbatim to provide guidance as to the type of characteristics of material that might be encountered in future recovery operations.

a. Initial analysis of the debris from the crash site seems to indicate that the debris is that of an extraterrestrial craft which exploded from within and came into contact with the ground with great force, completely destroying the craft. The volume of matter indicates that the craft was approximately the size of a medium aircraft, although the weight of the debris indicates that the craft was extremely light for its size.

b. Metallurgical analysis of the bulk of the debris recovered indicates that the samples are not composed of any materials currently known to Terrestrial science.

c. The material tested possesses great strength and resistance to heat in proportion to its weight and size, being stronger by far than any materials used in military or civilian aircraft at present.

d. Much of the material, having the appearance of aluminum foil or aluminum-magnesium sheeting, displays none of the characteristics of either metal, resembling instead some kind of unknown plastic-like material.

e. Solid structures and substantial beams having a distinct similarity in appearance to very dense grain-free wood, was very light in weight and possesses tensile and compression strength not obtainable by any means known to modern industry.

f. None of the material tested displayed measurable magnetic characteristics or residual radiation.

g. Several samples were engraved or embossed with marks and patterns. These patterns were not readily identifiable and attempts to decipher their meaning has been largely unsuccessful.

h. Examination of several apparent mechanical devices, gears, etc. revealed little or nothing of their functions or methods of manufacture.

MJ—12 4838B   6

LEON BIBI

TOP SECRET/MAJIC EYES ONLY

CHAPTER 3
RECOVERY OPERATIONS

Section I. SECURITY

12. Press Blackout

Great care must be taken to preserve the security of any location where Extraterrestrial Technology might be retrievable for scientific study. Extreme measures must be taken to protect and preserve any material or craft from discovery, examination, or removal by civilian agencies or individuals of the general public. It is therefore recommended that a total press blackout be initiated whenever possible. If this course of action should not prove feasible, the following cover stories are suggested for release to the press. The officer in charge will act quickly to select the cover story that best fits the situation. It should be remembered when selecting a cover story that official policy regarding UFOBs is that they do not exist.

a. Official Denial. The most desirable response would be that nothing unusual has occurred. By stating that the government has no knowledge of the event, further investigation by the public press may be forestalled.

b. Discredit Witnesses. If at all possible, witnesses will be held incommunicado until the extent of their knowledge and involvement can be determined. Witnesses will be discouraged from talking about what they have seen, and intimidation may be necessary to ensure their cooperation. If witnesses have already contacted the press, it will be necessary to discredit their stories. This can best be done by the assertion that they have either misinterpreted natural events, are the victims of hysteria or hallucinations, or are the perpetrators of hoaxes.

c. Deceptive Statements. It may become necessary to issue false statements to preserve the security of the site. Meteors, downed satellites, weather balloons, and military aircraft are all acceptable alternatives, although in the case of the downed military aircraft statement care should be exercised not to suggest that the aircraft might be experimental or secret, as this might arouse more curiosity of both the American and the foreign press. Statement issues concerning contamination of the area due to toxic spills from trucks or railroad tankers can also serve to keep unauthorized or undesirable personnel away from the area.

13. Secure the Area

The area must be secured as rapidly as possible to keep unauthorized personnel from infiltrating the site. The officer in charge will set up a perimeter and establish a command post inside the perimeter. Personnel allowed on the site will be kept to the absolute minimum necessary to prepare the craft or debris for transport, and will consist of Military Security Teams.

Local authorities may be pressed into service as traffic and crowd control. *Under no circumstances* will local official or law enforcement personnel be allowed inside the perimeter and all necessary precautions should be taken to ensure that they do not interfere with the operation.

a. Perimeter. It is desirable that sufficient military personnel be utilized to set up a perimeter around the site large enough to keep both unauthorized personnel and the

MJ—12 4838B

7

This unclassified copy is for research purposes. Rev 3; 2/26/98; Helvetica headers, Times text

TOP SECRET/MAJIC EYES ONLY

perimeter personnel from seeing the site. Once the site is contained, regular patrols will be set up along the perimeter to ensure complete security, and electronic surveillance will be utilized to augment the patrols. Perimeter personnel will be equipped with hand communication and automatic weapons with live ammunition. Personnel working at the site will carry sidearms. No unauthorized personnel will be allowed into the secured area.

  b. *Command Post.* Ideally, the command post should be as close to the site as is practical to efficiently coordinate operations. As soon as the command post is operational, contact with the Majestic—12 Group will be established via secure communications.

  c. *Area Sweep.* The site and the surrounding area will be cleared of all unauthorized personnel. Witnesses will be debriefed and detained for further evaluation by MJ—12. *Under no circumstances* will witnesses be released from custody until their stories have been evaluated by MJ—12 and they have been thoroughly debriefed.

  d. *Situation Evaluation.* A preliminary evaluation of the situation will be completed and a preliminary report prepared. The MJ—12 Group will then be briefed on the situation at the earliest possible opportunity. The MJ—12 Group will then make a determination as to whether or not a MJ—12 RED TEAM or OPNAC Team will be dispatched to the area.

Section II. TECHNOLOGY RECOVERY

14. Removal and Transport

  As soon as communication is established, removal and transport of all material will commence under order of MJ—12.

  a. *Documentation.* If the situation permits, care should be taken to document the area with photographs before anything is moved. The area will be checked for radiation and other toxic agents. If the area cannot be kept secure for an extended period of time, all material must be packed and transported as quickly as possible to the nearest secure military facility. This will be accomplished by covered transport using little-traveled roads wherever possible.

  b. *Complete or Functional Craft.* Craft are to be approached with extreme caution if they appear functional, as serious injury may result from exposure to radiation and electrical discharges. If the craft is functioning, but appears to be abandoned, it may be approached only by specially trained MJ—12 RED TEAM personnel wearing protective clothing. Any craft that appears to be functioning should also be left to MJ—12 RED TEAM disposal. Complete craft and parts of crafts too large to be transported by covered transport will be disassembled, if this can be accomplished easily and quickly. If they must be transported whole, or on open flatbed trailers, they will be covered in such a manner as to camouflage their shape.

  c. *Extraterrestrial Biological Entities.* EBEs must be removed to a top security facility as quickly as possible. Great care should be taken to prevent possible contamination by alien biological agents. Dead EBEs should be packed in ice at the earliest opportunity to preserve tissues. Should live EBEs be encountered, they should be taken into custody and removed to a top security facility by ambulance. Every effort should be taken to ensure the EBEs survival. Personnel involvement with EBEs alive or dead must be kept to an absolute minimum. (See Chapter 5 for more detailed instruction on dealing with EBEs.)

MJ—12 4838B   8

LEON BIBI

TOP SECRET/MAJIC EYES ONLY

15. Cleansing the Area

Once all material has been removed from the central area, the surrounding area will be thoroughly inspected to make sure that all traces of Extraterrestrial Technology have been removed. In the case of a crash, the surrounding area will be thoroughly gone over several times to ensure that nothing has been overlooked. The search area involved may vary depending on local conditions, at the discretion of the officer in charge. When the officer in charge is satisfied that no further evidence of the event remains at the site, it may be evacuated.

16. Special or Unusual Circumstances

The possibility exists that extraterrestrial craft may land or crash in heavily populated areas, where security cannot be maintained effectively. Large segments of the population and the public press may witness these craft. Contingency Plan MJ-1949-04P/78 (TOP SECRET * EYES ONLY) should be held in readiness should the need to make a public disclosure become necessary.

17. Extraterrestrial Technology   *(See Table on next page)*

Figure 2. MJ Form 1—007

ADAM = ALIEN

This unclassified copy is for research purposes. Rev 3; 2/28/98, Helvetica headers, Times text

TOP SECRET/MAJIC EYES ONLY

17. Extraterrestrial Technology Classification Table

No.	Item	Description or condition	MJ—12 Code	Receiving Facility
1	Aircraft	Intact, operational, or semi-intact aircraft of Extraterrestrial design and manufacture.	UA-002-6	Area 51 S-4
2	Intact device	Any mechanical or electronic device or machine which appears to be undamaged and functional.	ID-301-F	Area 51 S-4
3	Damaged device	Any mechanical or electronic device or machine which appears to be damaged but mostly complete.	DD-303N	Area 51 S-4
4	Powerplant	Devices and machines which are possible propulsion units, fuel and associated control devices and panels.	PD-40-8G	Area 51 S-4
5	Identified fragments	Fragments composed of elements or materials easily recognized as known to current science and technology, i.e., aluminum, magnesium, plastic, etc.	IF-101-K	Area 51 S-4
6	Unidentified fragments	Fragments composed of elements or materials not known to current science and technology and which exhibits unusual or extraordinary characteristics.	UF-103-M	Area 51 S-4
7	Supplies and provisions	Non-mechanical or non-electronic materials of a support nature such as clothing, personal belongings, organic ingestibles, etc.	SP-331	Blue Lab WP-61
8	Living entity*	Living non-human organisms in apparent good or reasonable health.	EBE-010	OPNAC BBS-01
9	Non-living entity	Deceased non-human organisms or portions of organisms, organic remains and other suspect matter.	EBE-XO	Blue Lab WP-61
10	Media	Printed matter, electronic recordings, maps, charts, photographs and film.	MM-54A	Building 21 KB-88
11	Weapons	Any device or portion of a device thought to be offensive or defensive weaponry.	WW-010	Area 51 S-4

*Living entity must be contained in total pending arrival of OPNAC personnel

MJ—12 4838B

TOP SECRET/MAJIC EYES ONLY
REPRODUCTION IN ANY FORM IS FORBIDDEN BY FEDERAL LAW

10

LEON BIBI

CHAPTER 4

RECEIVING AND HANDLING

Section I. HANDLING UPON RECEIPT OF MATERIAL

20. Uncrating, Unpacking, and Checking

(fig. 3)

Note. The uncrating, unpacking, and checking procedure for containers marked "MAJIC—12 ACCESS ONLY" will be carried out by personnel with MJ—12 clearance. Containers marked in this manner will be placed in storage in a top security area until such time as authorized personnel are available for these procedures.

a. Be very careful when uncrating and unpacking the material. Avoid thrusting tools into the interior of the shipping container. Do not damage the packaging materials any more than is absolutely necessary to remove the specimens; these materials may be required for future packaging. Store the interior packaging material within the shipping container. When uncrating and unpacking the specimens, follow the procedure given in (1) through (11) below:

    (1) Unpack the specimens in a top security area to prevent access of unauthorized personnel.

    (2) Cut the metal wires with a suitable cutting tool, or twist them with pliers until the straps crystallize and break.

    (3) Remove screws from the top of the shipping container with a screw driver.

    (4) Cut the tape and seals of the case liner so that the waterproof paper will be damaged as little as possible.

    (5) Lift out the packaged specimens from the wooden case.

    (6) Cut the tape which seals the top flaps of the outer cartons; be careful not to damage the cartons.

    (7) Cut the barrier along the top heat sealed seam and carefully remove the inner carton.

    (8) Remove the sealed manila envelope from the top of the inner carton.

    (9) Open the inner carton and remove the fiberboard inserts, dessicant, and humidity indicator.

    (10) Lift out the heat sealed packaging containing the specimens; arrange them in an orderly manner for inspection.

    (11) Place all packaging material in the shipping container for use in future repacking.

[In place here is Figure 3—a diagram showing how to package non-organic extraterrestrial specimens.]

b. Thoroughly check all items against the shipping documents. Carefully inspect all items for possible damage during shipping or handling. Sort the items according to classification number in preparation for transfer to the designated Laboratory or department. Laboratory or department personnel are responsible for transporting items to the designated areas. This will be accomplished as quickly as possible by covered transport escorted by security personnel.

MJ—12 4838B   12

This unclassified copy is for research purposes. Rev 3; 2/28/06; Helvetica headers, Times text
TOP SECRET/MAJIC EYES ONLY

CHAPTER 5

EXTRATERRESTRIAL BIOLOGICAL ENTITIES

Section I. LIVING ORGANISMS

21. Scope

a. This section deals with encounters with living Extraterrestrial Biological Entities (EBEs). Such encounters fall under the jurisdiction of MJ-12 OPNAC BBS—01 and will be dealt with by this special unit only. This section details the responsibilities of persons or units making the initial contact.

22. General

Any encounter with entities known to be of extraterrestrial origin is to be considered to be a matter of national security and therefore classified TOP SECRET. Under no circumstance is the general public or the public press to learn of the existence of these entities. The official government policy is that such creatures do not exist, and that no agency of the federal government is now engaged in any study of extraterrestrials or their artifacts. Any deviation from this stated policy is absolutely forbidden.

23. Encounters

Encounters with EBEs may be classified according to one of the following categories:

a. Encounters initiated by EBEs. Possible contact may take place as a result of overtures by the entities themselves. In these instances it is anticipated that encounters will take place at military installations or other obscure locations selected by mutual agreement. Such meeting would have the advantage of being limited to personnel with appropriate clearance, away from public scrutiny. Although it is not considered very probable, there also exists the possibility that EBEs may land in public places without prior notice. In this case the OPNAC Team will formulate cover stories for the press and prepare briefings for the President and the Chiefs of Staff.

b. Encounters as the result of downed craft. Contact with survivors of accidents or craft downed by natural events or military action may occur with little or no warning. In these cases, it is important that the initial contact be limited to military personnel to preserve security. Civilian witnesses to the area will be detained and debriefed by MJ-12. Contact with EBEs by military personnel not having MJ-12 or OPNAC clearance is to be strictly limited to action necessary to ensure the availability of the EBEs for study by the OPNAC Team.

24. Isolation and Custody

a. EBEs will be detained by whatever means are necessary and removed to a secure location as soon as possible. Precautions will be taken by personnel coming in contact with EBEs to minimize the risk of disease as a result of contamination by unknown organisms. If the entities are wearing space suits or breathing apparatus of some kind, care should be exercised to prevent damage to these devices. While all efforts should be taken to assure the well-being of the EBEs, they must be isolated from any contact with unauthorized personnel. While it is not clear what provisions or amenities might be

TOP SECRET/MAJIC EYES ONLY
REPRODUCTION IN ANY FORM IS FORBIDDEN BY FEDERAL LAW

˅227˅

TOP SECRET/MAJIC EYES ONLY

required by non-human entities, they should be provided if possible. The officer in charge of the operation will make these determinations, since no guidelines now exist to cover this area.

b. Injured or wounded entities will be treated by medical personnel assigned to the OPNAC Team. If the team medical personnel are not immediately available, First Aid will be administered by Medical Corps personnel at the initial site. Since little is known about EBE biological functions, aid will be confined to the stopping of bleeding, bandaging of wounds and splinting of broken limbs. No medications of any kind are to be administered as the effect of terrestrial medications on non-human biological systems are impossible to predict. As soon as the injuries are considered stabilized, the EBEs will be moved by closed ambulance or other suitable conveyance to a secure location.

c. In dealing with any living Extraterrestrial Biological Entity, security is of paramount importance. All other considerations are secondary. Although it is preferable to maintain the physical well-being of any entity, the loss of EBE life is considered acceptable if conditions or delays to preserve that life in any way compromises the security of the operations.

d. Once the OPNAC Team has taken custody of the EBEs, their care and transportation to designated facilities become the responsibility of OPNAC personnel. Every cooperation will be extended to the team in carrying out duties. OPNAC Team personnel will be given TOP PRIORITY at all times regardless of their apparent rank or status. No person has the authority to interfere with the OPNAC Team in the performance of its duties by special direction of the President of the United States.

Section II. NON-LIVING ORGANISMS

25. Scope

Ideally, retrieval for scientific study of cadavers and other biological remains will be carried out by medical personnel familiar with this type of procedure. Because of security considerations, such collection may need to be done by non-medical personnel. This section will provide guidance for retrieval, preservation, and removal of cadavers and remains in the field.

26. Retrieval and Preservation

a. The degree of decomposition of organic remains will vary depending on the length of time the remains have been lying in the open unprotected and may be accelerated by both local weather conditions and action by predators. Therefore, biological specimens will be removed from the crash site as quickly as possible to preserve the remains in as good a condition as possible. A photographic record will be made of all remains before they are removed from the site.

b. Personnel involved in this type of operation will take all reasonable precautions to minimize physical contact with the cadavers or remains being retrieved. Surgical gloves should be worn or, if they are not available, wool or leather gloves may be worn provided they are collected for decontamination immediately after use. Shovels and entrenching tools may be employed to handle remains provided caution is exercised to be certain no damage is done to the remains. Remains will be touched with bare hands only if no other means of moving them can be found. All personnel and equipment involved in recovery operations will undergo decontamination procedures immediately after those operations are [sic] have been completed.

MJ—12 4838B

TOP SECRET/MAJIC EYES ONLY

c. Remains will be preserved against further decomposition as equipment and conditions permit. Cadavers and remains will be bagged or securely wrapped in waterproof coverings. Tarpaulins or foul weather gear may be used for this purpose if necessary. Remains will be refrigerated or packed with ice if available. All remains will be tagged or labeled and the time and date recorded. Wrapped remains will be placed on stretchers or in sealed containers for immediate removal to a secure facility.

d. Small detached pieces and material scraped from solid surfaces will be put in jars or other small capped containers if available. Containers will be clearly marked as to their contents and the time and date recorded. Containers will be refrigerated or packed with ice as soon as possible and removed to a secure facility.

[In place here is Figure 4—diagrams of the various types of extraterrestrial craft discussed in the text.]

TOP SECRET/MAJIC EYES ONLY

TOP SECRET/MAJIC EYES ONLY

CHAPTER 6

GUIDE TO UFO IDENTIFICATION

Section I. UFOB GUIDE

27. Follow-up Investigations

A UFOB report is worthy of follow-up investigation when it contains information to suggest that positive identification with a well-known phenomenon may be made or when it characterizes an unusual phenomenon. The report should suggest almost immediately, largely by the coherency and clarity of the data, that there is something of identification and/or scientific value. In general, reports which should be given consideration are those which involve several reliable observers, together or separately, and which concern sighting of greater duration than one quarter minute. Exceptions should be made to this when circumstances attending the report are considered to be extraordinary. Special attention should be given to reports which give promise to a "fix" on the position and those reports involving unusual trajectories.

28. Rules of Thumb

Each UFOB case should be judged individually but there are a number of "rules of thumb," under each of the following headings, which should prove helpful for determining the necessity for follow-up investigation.

a. Duration of Sighting. When the duration of a sighting is less than 15 seconds, the probabilities are great that it is not worthy of follow-up. As a word of caution, however, should a large number of individual observers report an unusual sighting of a few seconds duration, it should not be dismissed.

b. Number of Persons Reporting the Sighting. Short duration sightings by single individuals are seldom worthy of follow-up. Two or three competent independent observations carry the weight of 10 or more simultaneous individual observations. As an example, 25 people at one spot may observe a strange light in the sky. This, however, has less weight than two reliable people observing the same light from different locations. In the latter case a position-fix is indicated.

c. Distance from Location of Sightings to Nearest Field Unit. Reports which meet the preliminary criterion stated above should all be investigated if their occurrence is in the immediate operating vicinity of the squadron concerned. For reports involving greater distances, follow-up necessity might be judged as being inversely proportional to the square of the distances concerned. For example, an occurrence 150 miles away might be con-

TOP SECRET/MAJIC EYES ONLY
REPRODUCTION IN ANY FORM IS FORBIDDEN BY FEDERAL LAW

This unclassified copy is for research purposes. Rev 3; 2/28/95; Helvetica headers, Times text

TOP SECRET/MAJIC EYES ONLY

APPENDIX I

REFERENCES

There is some writing here

No. 4, AB

---

1. [Applicable] Regulations

-4

Military security (Safeguarding Security Information.

Maintenance Supplies and Equipment, Maintenance Responsiblities and Shop Operation.

2. Supply

xx 725-405-5

Preparation and Submission of Requisitions for Supplies.

3. Other Publications

XX 219-20-3   Index of Training Manuals.

XX 310-20-4   Index of Technical Manuals, Technical Regulations, Technical Bulletins, Supply Bulletin, Lubrications Orders, and Modification Work Orders.

XX310-20-5   Index of Administrative Publications.

XX310-20-7   Index of Tables of Organization and Equipment, Reduction Tables, Tables of Organization, Tables of Equipment, Type Tables of Distribution and Tables of Allowance.

4. Test Equipment References

TM 11—664   Theory and Use of Electronic Test Equipment.

5. Photographic References

TM 11—404A   Photographic Print Processing Unit AN/TFQ-9.

TM 11—405   Processing Equipment PH—406.

TM 11—401   Elements of Signal Photography.

TM 11—2363   Darkroom PH—392.

MJ—12 4838B

17

TOP SECRET/MAJIC EYES ONLY
REPRODUCTION IN ANY FORM IS FORBIDDEN BY FEDERAL LAW

The Pascal Tomb – One of the greatest finds in
Archaeological history. This engraving made me a believer.
Take a long look at this carefully. This is a clear depiction of
a human being (Pascal), operating a rocket propulsion
vehicle. The following is shown:

Breathing apparatus in his nose
Right hand pushing button(s)
Left hand operating a lever
Right foot operating a break?
Left foot igniting a thrust
A seat
An engine behind the seat
Rocket fumes in back
Ammunition (bullets?) on front, bottom and top

CHAPTER 21
CONCLUSION

I wish I saw that UFO ... I really do. But at the end of the day, it was that interest that sparked a 3 year quest into the unknown. The UFO's led to the EBE's, which led to the Anunnaki, which led to our origins. I always wondered why there were so many holes in the story of the Bible. Now I know.

It's amazing how many people are truly interested in this subject. At least 50% of the people that I speak to are interested in, and believe in, extra-terrestrials. "There has to be" they all say "there are so many planets, so many galaxies" ... "it's a certainty." "but not here ... no way ... they never came here."

Therein lies the rub. My fellow readers, they did come here. They still come here. It's because of them that *we are here*.

When I discuss this topic with strangers on an airplane, or at a restaurant, most people are open-minded, up until the Anunnaki part. That's where they tend to squint their eyes, and say "Annu-what?" I know, the name sounds funny ... it sounds like a fairy tale. I thought about writing a children's book called "Annu and Naki" which told the story of Enlil and Enki, animated and easy to read. It would recant the tale of two brothers, jealous of one another and fighting for power, wealth and their parents' attention. A story within a story. They came from a planet that was dying because of their inhabitants' misuse of the planets' resources.

Gold was what would save their planet from annihilation.

Annu and Naki found the gold on earth. They toiled in the mines digging it out. And then they had an idea. Let's create some 'helpers'. The helpers will do all the dirty work and get us what we need to save our planet. But we can't make them too smart, because they will overthrow us, so will just make them smart enough to follow our commands. Boy did we make some mistakes along the way. But we finally did it! We made a living, breathing, human being who can think and speak.

The irony is that we humans are now, in the second millennium, doing the same thing that the Anunnaki did--we are destroying our resources, and depleting our ozone layer. It gets hotter every year. Radiation from our sun is at its peak. The poles are melting and water levels are rising. Maybe someday, when the ozone layer gets too thin, a brave crew of fifty humans will fly to another planet to cull it of its resources in an effort to save us.

I wonder what I would say to the nine foot tall, muscular, bearded Anunnaki 'god' that created us ... I would probably say--

"Thanks for giving me the opportunity to live, and breath and think. I thought a lot about you and what you would look like. And now that we have finally met, you really look a lot like me."

BIBLIOGRAPHY

Alexander, John B. *UFOs: Myths, Conspiracies, and Realities*. New York: St. Martin's Press, 2011.

Bauval, Robert, and Adrian Gilbert. *The Orion Mystery*. New York: Crown Publishing Group, 1994.

Bell, Kelly. *Visitors: A New Look At UFOs*. Lincoln, NE: iUniverse, 2007.

Bennett, Jefferey. *Beyond UFOs*. Princeton, NJ: Princeton University Press, 2008.

Bennet, Mary, and David S. Percy. *Dark Moon*. Kempton, IL: Adventures Unlimited Press, 2001.

Brennan, Herbie. *Martian Genesis*. New York: Dell, 1998.

Bruce, Alexandra. *The Philadelphia Experiment Murder*. Westbury, NY: Sky Books 2001.

Carey, Thomas J., and Donald R. Schmitt. *Witness to Roswell*. Franklin Lakes, NJ: Career Press, 2007.

Cayce, Edgar Evans. *Edgar Cayce on Atlantis*. New York: Grand Central, 1968.

Childress, David Hatcher. *Technology of the Gods: The Incredible Science of the Ancients*. Kempton IL: Adventures Unlimited Press, 2000.

Childress, David Hatcher. *The Time Travel Handbook*. Kempton, IL: Adventures Unlimited Press, 1999.

Collins, Robert M., and Richard C. Doty. *The Black World of UFOs: Exempt from Disclosure*. Vandalia, OH: Peregrine Communications, 2008.

Conway, Patrick J. *The Awful Truth*. Bellarose, NY: Pragmatic Press, 2000.

Cooper, William. *Behold a Pale Horse*. Flagstaff, AZ: Light Technology, 1991.

Coppens, Phillip, *The Ancient Alien Question*. Pompton Plains, NJ: Career Press, 2012.

Corso, Philip J. *The Day After Roswell*. New York: Pock Nooks, 1997.

Davis, Scott P. *Our True Origin*.

De Lafayette, Maximillien. *1520 Things You Don't Know About Ancient Aliens, UFOs, Alien Technology, Extraterrestrials Black Operations, Part 2*. New York: Times Square Press, 2011.

Dolan, Richard M. *UFOs and the National Security State*. Charlottesville, VA: Hampton Roads, 2002.

Farrell, Joseph P. *Genes, Giants, Monsters, and Men*. Port Townsend, WA: Feral House, 2011.

Farrell, Joseph P. and Scott D. de Hart. *The Grid of the Gods*. Kempton, IL: Adventures Limited Press, 2011.

Freidman, Stanton T. *Flying Saucers and Science*. Franklin Lakes, NJ: Career Press, 2008.

Freidman, Stanton T. *Top Secret/Majic*. New York: Marlowe, 1996.

Haas, George J. and William R. Saunders. *The Cydonia Codex*. Berkeley, CA: Frog, 2005.

Hart, Will. *The Genesis Race*. Rochester, Vermont: Bear, 2003.

Hesemann, Michael, and Philip Mantle. *Beyond Roswell*. London: Michael O'Mara Books, 1997.

Jones, Marie D. and Larry Flaxman. *This Book is from the Future*. Pompton Plains, NJ: Career Press, 2012.

Kasten, Len. *The Secret History of Extraterrestrials*. Rochester, VT: Bear, 2010

Kean, Leslie. *UFOs: Generals, Pilots, and Government Officials Go on the Record*. New York: Harmony Books, 2010.

Keith, Jim. *Mind Control and UFOs: Casebook on Alternative 3*. Kempton, IL: Adventures Unlimited Press, 2005.

Kenyon, J. Douglas. *Forbidden Science*. Rochester, VT: Bear, 2008.

Komarek, Ed. *Ufos, Exopolitics, and The New World Order*. Cairo, Ga: Shoestring, 2012.

Korff, Kal K. *The Roswell UFO Crash*. New York: Dell, 2000.

Kreisberg, Glenn. *Lost Knowledge of the Ancients*. Rochester, VT: Bear, 2010.

LaViolette, Paul A. *Secrets of Antigravity Propulsion*. Rochester, VT: Bear, 2008.

Marcel, Jesse Jr. and Linda Marcel. *The Roswell Legacy*. Franklin Lakes, NJ: Career Press, 2009.

McDowell, Al. *Uncommon Knowledge*. Bloomington, IN: Author House, 2009.

Nichols, Preston B. and Peter Moon. *Pyramids of Montauk*. Westbury, NY: Sky Books, 1998.

O'Neil, John J. *Prodigal Genius: The Life of Nikola* Tesla. Kempton, IL. Adventures Unlimited Press, 2008.

Patton, Phil. *Dreamland*. New York: Random House, 1998.

Pye. Lloyd. Everything *You Know Is Wrong, Book One: Human Origins*. Lincoln, NE: iUniverse, 1997.

Pye, Michael, and Kirsten Dalley. *Lost Civilizations and Secrets of the Past*. Pompton Plains, NJ: Career Press, 2012.

Redfern, Nick. *The Pyramids and the Pentagon*. Pompton Plains, NJ: Career Press, 2012.

Roberts, Scott Alan. *The Rise and Fall of Nephilim*. Pompton Plains, NJ: Career Press, 2012.

Rutkowski, Chris A. *A World of UFOs*. Ontario, Canada: Dundurn Press, 2008.

Sitchin, Zecharia. *Journeys to the Mystical Past*. Rochester, VT: Bear, 2007.

Steckling, Fred. *We Discovered Alien Bases on the Moon*. Vista, CA: GAF International, 1981.

Steiger, Brad. *The Philadelphia Experiment and Other UFO Conspiracies*. New Brunswick, NJ: Inner Light, 1990.

Tellinger, Michael. *Slave Species of the Gods*. Rochester, VT: Bear, 2012.

Temple, Robert. *The Sirius Mystery*. Rochester, VT: Destiny Books, 1998.

Thomas, Robert Steven. *Intelligent Intervention*. Indianapolis, IN: Dog Ear, 2011.

Von Daniken, Erich. *The Eyes of the Sphinx*. New York: Berkley,1996.

Von Ward, Paul. *We've Never Been Alone*. Charlottesville, VA: Hampton Roads, 2011.

Wells, K.B. Jr. *The Montauk Files*. Tempe, AZ: New Falcon, 2007.

Wilkinson, Frank G. *The Golden Age of Flying Saucers*. New Paradigm Press, 2007.

AFTERWORD

PEOPLE HAVE OFTEN ASKED ME when I would be writing a follow up to *"Adam = Alien"*, and I'd always respond "soon, soon, probably next year". Shortly after *"Adam = Alien"* was published I received a flurry of invitations to speak at Colleges, Universities, UFO Conferences, even at local restaurants whose owners were interested in the subject matter. I sold my business of 25 years, changed occupations and then recently, lost my father who died at 94. He was a decorated WW11 veteran fighting in Europe from 1941-1945. Whenever I asked him if he'd ever seen or heard about Foo Fighters during the War, he surprisingly said, "No, not really".

A supremely humble man, who was frugal with words, was as honest as they come - a true gentleman.
I couldn't believe that he had never seen or heard about Foo Fighters, when thousands of veterans swore by them. Maybe since he had been with a platoon of land-based amphibious engineers from Brooklyn, NY, they had not been exposed to airborne Foo Fighters like the Air Force pilots would have been. But you would think that he'd at least heard about them. "No, not really".
I hear "no, not really" quite often when I ask people if they believe in UFO's, would consider alternative truths to the Bible, or believe that extraterrestrials have indeed landed here on Earth - even in the U.S! But it doesn't deter me. I keep arguing, fighting, coaxing, prodding away with more and more and more incontrovertible

evidence to support my thesis (and others shared by so many notable authors in the Ancient Alien space). **I believe in it so strongly that it has become fact.** I believe in it so strongly that I changed 180 degrees from a religious, God-fearing boy to an agnostic non-God-fearing adult, to a secular atheist who only attends religious ceremonies out of respect for others' rights, beliefs and freedoms. I am now an atheist, who believes in a master architect in the universe - just not the one that the Bible purports to be the only God of the Bible. Yes, there is an elegant grand design, beautifully constructed through complex mathematics that pervades from the spiral helix of our DNA strands to the deepest abyss of a Black Hole.

If you, my dear reader, could believe just 10% of what I have written, then I have succeeded. I hope that I have enlightened you just bit, the same way that authors like Sitchin, Pye, Alford, and Von Daniken have truly inspired me. There are so many secrets out there. So many people with so many agendas. So many people who blindly follow other people's direction. So may people who truly believe that world governmental leaders have their best interest at heart. So many people who believe that only what they are taught in school is the truth.

So many people who believe every word in the Bible. Am I one of those people?

No, not really.

ABOUT THE AUTHOR

LEON BIBI is the author of THE ADAM SERIES.

Leon is a historian and avid researcher on all subjects regarding archaeology, human and ancient history, biology, Egyptology and religion. Leon has been a featured guest speaker on George Noory's "Coast to Coast" radio show, The Conspiracy Show with Richard Syrett, Revolution Radio, and will be on the Ancient Aliens Series of the History Channel next year. He has been invited to speak about the Ancient Alien subject at various conferences, and both radio and television. Leon has a B.A from Washington University in St. Louis and attended Law School at the University of Miami (FL). He was a CEO of a Consumer Products company for 22 years, and a member of the prestigious "Young Presidents Organization" for 17 years. He is a professional drummer playing and recording music throughout his adult life. Leon lives with his family on the East Coast of the United States.

BOOK 2 OF THE ADAM
TRILOGY

This book, the second in the Adam Series, exposes a 4,000 year old secret. It involves the reader to try and decode elements of the Sumerian tablets, the Bible and ancient forbidden texts It is a controversial and explosive expose on the evidence given in plain site to support the impact that extraterrestrials have had on the human experience. Adam Decoded uses code written within the words of these ancient texts to decipher the true meaning of human evolution.

What is Darwin's "missing link"? Why can't fossils be found to support his claim that homo sapiens evolved directly (and only) from primates in the natural course of time? Is the Bible an accurate, fact-based, history of man's true origins?

Book 2 will unveil all of this and more.

ON SALE NOW AT

www.amazon.com

BOOK 3 OF THE ADAM TRILOGY

COMING SOON!

Coming Soon!

This book will delve into blood and chromosomal evidence of extraterrestrial tampering and intervention in the creation of modern day homo sapiens. It will explore the motives and actions of the Anunnaki gods who used their own DNA to create the world's first human hybrid hundreds of thousands of years ago. It will demonstrate that the supposed mythology of the Sumerian tablets which clearly told its story was a true record of our origins. How the Old Testament and New Testament was a recapitulation of the tablets yet altered to serve a purpose. This purpose was self-serving.

The truth lies in these tablets, spoken without an agenda, recounting the story of the creation of the Earth and all multicellular organisms created not solely through Survival of the Fittest and Natural Selection. We had help from the gods.

How did they do it? What secrets did they bestow upon us? What level of intelligence can we reach?

Book 3 will unveil all of this and more.

The Adam Code continues...

ADAM
=ALIEN

VOL. 1.2

⋊ LEON BIBI ⋈

Made in the USA
Las Vegas, NV
12 June 2021